Excel 2021

A Complete Guide To Master Excel's Fundamentals And Understanding Of Advanced Features Such As Sampling Technique, Business Modelling And Numerous Methods Of Data Analysis

JIAYI SIMONDS

Contents

Introduction

Microsoft Excel is a spreadsheet application developed by Microsoft that enables users to coordinate, format, and analyze information using formulas. This package is used in the Microsoft Office suite and is consistent with most Office programs. As with other Microsoft Office services, Microsoft Excel could also be purchased in the cloud through Office 365. Excel operates in the same way as a database, with formulas, records, and functions organized in columns (represented by letters) and the rows (represented by numbers) that can be used to conduct complex calculations. Microsoft released the first version of Excel in 1985, and by the 1990s, it had become one of the world's most widely used and significant computing tools. Excel is also a commonly popular technology installed on almost every private and professional computer on the planet. Simply, Excel is the most convenient tool for organizing and managing financial records, which is why so many companies use it. It enables enough customization and adaptability in its application.

Another reason to use Excel is the simplicity of use. With little to no experience or training, a user can open a workbook, begin inputting data, and review and analyze data. Microsoft Excel is an extremely useful and versatile data collection and documentation tool. It is a spreadsheet application that includes columns and rows, with each intersection of the column and a row being referred to as a "cell." Each cell contains a single piece of data or information. By grouping the data in this manner, you will easily locate information and quickly extract information from evolving data. MS Excel enables consumers to organize data to examine multiple variables from a variety of perspectives. Microsoft Visual Basic is a programming language for Excel applications that enables users to develop a multitude of advanced numerical approaches. Programmers can write code directly in the Visual Basic Editor, with Windows, for writing, debugging, and organizing code modules.

Microsoft Excel is unquestionably a difficult program to understand and use. This is why it is often preferable to allow others to assist us with understanding how to deal effectively with it. Indeed, regardless of whether you are a:

- A student wishing to understand how to use Microsoft Excel in order to complete a school assignment.
- Interested in learning Microsoft Excel for business purposes.
- Interested in expanding your experience and acquiring new skills
- Have a basic understanding of Microsoft Excel for personal use

Or for some other reasons why this book is sufficient for you. By reading this Microsoft Excel guide, you can access numerous high-quality illustrations, tips, and tricks. Additionally, you will have a thorough guide to all Excel fundamentals, enabling you to regularly interact more confidently in Microsoft Excel. Therefore, you must give this book a chance for it provides all the necessary information.

Chapter No: 1 Introduction To Microsoft Excel

1.1 What is Microsoft Excel?

Microsoft Excel is a spreadsheet tool that allows you to track and analyze mathematical and numerical data. Microsoft Excel comes with a variety of methods for conducting different functions, including macro programming, equations, pivot tables, and graphing tools. It is compatible with several different operating systems, Mac OS X, Android, including Windows, and iOS.

In an Excel spreadsheet, a table is generated using a sequence of columns and rows. Typically, columns are given alphabetical letters, while rows are assigned numbers. A cell is a point where a column and a row meet. The address of a cell is defined by the row number and the column letter.

Microsoft Excel is a database tool that is used as part of the Microsoft Office range of products. Spreadsheets contain values organized in rows and columns, which can be mathematically modified using simple and complex arithmetic operations. Along with the usual spreadsheet functionality, Excel includes support for programming via MS visual basic for the applications (VBA), as well as the ability to view data from the outside sources via MS dynamic data exchange (DDE). Excel is a spreadsheet program of computer.

Microsoft Excel was first introduced in 1985 for Macintosh computers, supplemented by a Windows version in 1987. Check out the following list to learn about the latest Excel updates for Windows:

1. Excel 2.0, (1987)
2. Excel 3.0, (1990)
3. Excel 4.0, (1992) Included in the Microsoft Office 3.0
4. Excel 5.0, (1993) Included in the Microsoft Office 4.0
5. Excel 95, (1995) Included in the Microsoft Office 95
6. Excel 97, (1997) Included in the Microsoft Office 97
7. Excel 2000, (2000) Included in the Microsoft Office 2000
8. Excel 2002, (2002) Included in the Microsoft Office XP
9. Excel 2003, (2003) Included in the Microsoft Office 2003
10. Excel 2007, (2007) Included in the Microsoft Office 2007
11. Excel 2010, (2010) Included in the Microsoft Office 2010
12. Excel 2013, (2013) Included in the Microsoft Office 2013

1.2 History of MS Excel

Microsoft Excel was a vital component of corporate bookkeeping and record-keeping in the early days of open PC business computing. A table with the autosum format is one of the better instances of an MS Excel use case. It is quite simple to insert a column of values in Microsoft Excel, press into a cell at the bottom of the spreadsheet, and then click the "autosum" button to have that cell add up all of the values entered above. This eliminates the need for manual ledger counts, a time-consuming aspect of business prior to the evolution of the modern spreadsheet. MS Excel is a must-have for different types of corporate programming, including analyzing regular, weekly, or monthly numbers, tabulating payroll and taxation, and performing other related business processes. Microsoft Excel is a spreadsheet program developed by the Microsoft Corporation in 1985. Excel is a widely used database program that organizes data in columns and rows and can be modified using formulas to execute mathematical operations on the data.

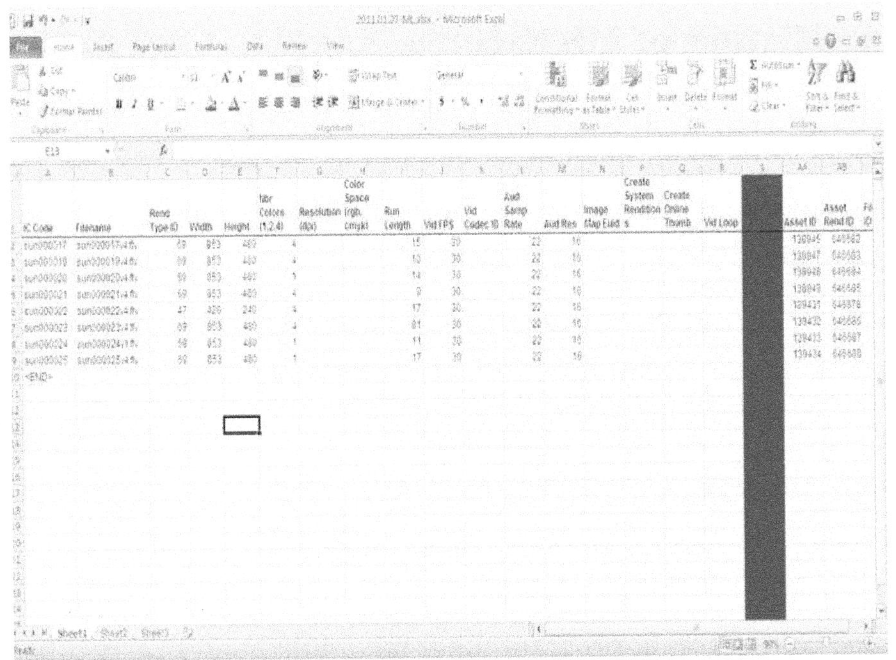

Lotus 1-2-3, introduced in 1982 by the Lotus development corporation, dominated the mid-1980 spreadsheet market for the personal computers (PCs) running Microsoft's MS-DOS operating system. Microsoft created a rival calculator, Excel, which was launched in 1985 for Apple Inc.'s Macintosh device. The new application rapidly gained popularity due to its powerful visuals and fast processing. Lotus 1-2-3 was not available on the Macintosh, allowing Excel to build a following among Mac users. Excel's next version, and the first to run on Microsoft's modern Windows operating system, was released in 1987. The influential software gained popularity due to its graphics-heavy interface, which was optimized to operate on the new Windows computers. Due to Lotus's delay in releasing a Windows implementation of its database, Excel gained market share and gradually became the leading spreadsheet technology in the mid-1990s. Later iterations of Excel saw major enhancements such as outlining, toolbars, drawing, three-dimensional maps, various shortcuts, and increased automation. In 1995, Microsoft modified Excel's naming convention to emphasize the product's initial year of availability. Excel 95 was built for the new 32-bit Intel Corporation 386 microprocessor-based computers. New versions were released in 1997 (Excel 97) and 1999 (Excel 1999). (Excel 2000). Excel 2002 was launched in 2003 as part of the Office XP suite. It included a major new function that allowed users to restore Excel data in the case of a computer crash.

Excel 2007 introduced a new user interface that integrated functionality from Microsoft's Word and PowerPoint systems, enabling users to switch between these applications seamlessly. Additionally, the development of charts, data exchange, security, sorting, formula writing, and filtering have been enhanced. Numerous basic usage cases elevated Microsoft Excel to critical end-user technology, valuable for training and professional growth. MS Excel has been used in basic business diploma programs on business computing for many years, and temporary employment agencies can test individuals on their ability to use Microsoft Word and Microsoft Excel to perform a variety of clerical tasks.

However, Microsoft Excel has become largely obsolete in several respects as enterprise technology has progressed. This is due to a term referred to as "visual dashboard" technology or "data visualization." In general, businesses and providers have developed innovative new approaches to visually display data that do not need end consumers to view a conventional spreadsheet of columns of numbers and identifiers. Other than that, they examine diagrams, charts, and other sophisticated displays in order to further and more easily comprehend the statistics. Individuals also recognized that visual presentations are significantly simpler to "interpret." The data visualization principle has altered the usage cases for Microsoft Excel. Whereas companies once used Microsoft Excel to manage hundreds of documents, most company usage cases currently include spreadsheets that manage little more than a few dozen values on each given project. If the spreadsheet contains more than a couple of hundred rows, the data would be more effectively displayed on a visual dashboard than on a conventional spreadsheet format.

1.3 Features of Microsoft Excel

1. **Effortless spreadsheet creation:** Create and manage spreadsheets of every scale, from personal or school records to government databases.

2. **Extensive toolset**: Take use of the most specialized features included in current spreadsheet applications,

such as advanced formulas, lookup formulas, and pivot tables, maps, sorting and filtering results, conditional formatting, data explorer, structural references, data interpretation, VBA, macros, and automation.

3. **Analytics:** Use built-in analytics tools to gain actionable knowledge.
4. **Audit Trail:** Using the integrated Review and Track Changes tools, keep track of changes to audit reports.
5. **Advanced calculations**: Raise the level for financial and data processing.
6. **Charting:** Visualize data using a variety of charts which can be exported to other Office applications.
7. **Multi-user collaboration:** The Office 365 integration allows all Excel customers to exchange and work on documents conveniently.
8. **Templates:** Simplify document creation with the help of customizable and powerful template offers.

1.4 How to open Microsoft excel?

The primary goal of this book is to educate you not just about the fundamentals of Microsoft Excel but also about how to master pivot tables, macros, formulas, VBA, and data processing and how to secure your first work as an Excel expert. Excel is a Windows program that can be managed in the same manner as the other Windows application. If you are using Windows for a graphical user interface, follow the instructions below (Windows XP, Vista or 7).

1. To start Microsoft Excel 2021, navigate to the Windows Start menu and choose Start All Programs. After that, choose Microsoft Office and then Microsoft Excel 2021.
2. (If you do not have Excel, type it after choosing the tab from the start menu.) A new blank workbook has been created and is waiting for you to complete the details.

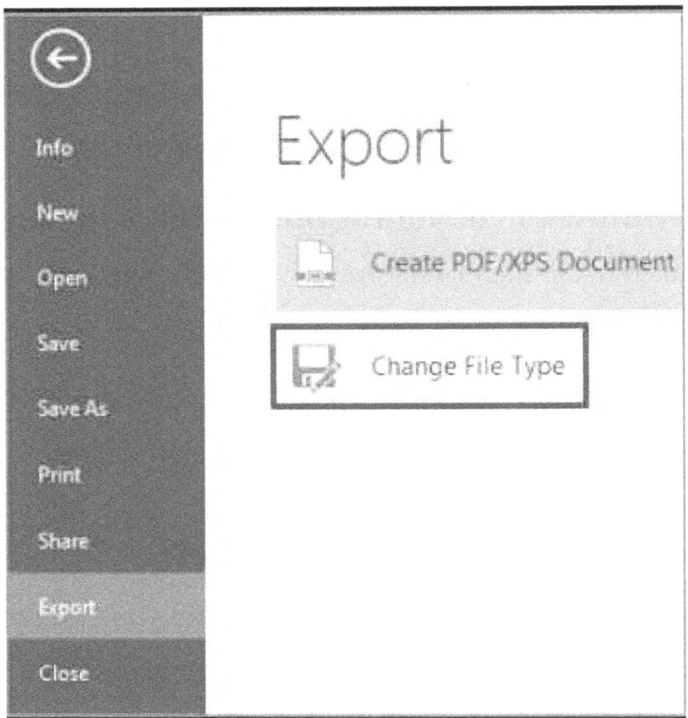

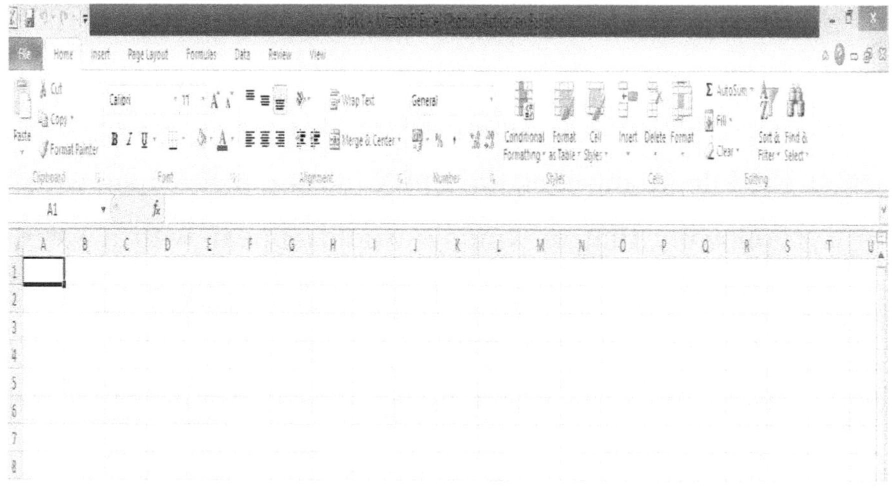

If you are attempting to access a previously saved file, you can first follow these instructions.

1. **Step 1:** Double-click any XLSX file to open it. Double-click the File in Microsoft Excel to access XLSX files. If you have already downloaded and installed a version of Microsoft Excel on your device (2016 or higher), double-clicking on a file would automatically open.
2. **Step 2:** Open Excel and begin dragging and dropping files inside it. If your computer already has Microsoft Excel loaded, you can simply drag the XLSX file into the open spreadsheet in Microsoft Excel. To do so, select the XLSX file with the left mouse button held down, move it further into an already-opened Excel spreadsheet, and then release your mouse button. After that, the XLSX sort file will be enabled.

3. **Step 3**: From the right-click option, choose "Open with." If you have a version of Microsoft Excel installed on your device, the pop-up menu would also enable you to access the File. Right-click the XLSX file and pick "Open with." Following that, a window would appear with a list of compatible programs with the File in question. When you click from Ms. Excel, the above File will open. If Microsoft Excel is not listed, your computer does not have it installed.

1.5 The Function library

A function is a predefined formula that conducts calculations with predefined values in a predefined order. Excel provides many useful functions for easily calculating the number, average, maximum value, count, and minimum value of a set of cells. To use functions properly, you must first consider the various components of a function and build arguments for calculating values and cell references.

The parts of a function

To program properly, a function should be written in a certain manner, referred to as the syntax. A function's simple syntax consists of an equals sign (=), the function's name (for example, SUM), and one or more reasons. Arguments include the data to be calculated. In the illustration below, the method will add the values in the cell range A1:A20.

Creating a function

Excel has a range of features. The following are some of the more often used functions:

1. **SUM:** This feature sums all of the values in the argument's cells.
2. **AVERAGE:** This method returns the average of the argument's values. It computes the total of the cells and then divides the result by the argument's cell count.
3. **COUNT**: This function counts the number of cells in the argument that contain numerical details. This role is useful for easily calculating the number of items included within a cell range.
4. **MAX:** This function determines the argument's maximum cell value.

5. **MIN:** This function determines the argument's lowest cell value.

Although Excel contains hundreds of functions, the ones you'll need the most will rely on the type of data included in your workbooks. There is no need to master any function, but being familiar with some of the more common forms of functions can assist you in developing new ventures. Additionally, you can use the Formulas tab's Feature Library to search functions by types, such as Logical, Text, Financial, and Date & Time.

1. To use the Function Library on the Ribbon, click the Formulas tab. Look for the group Function Library.
2. To read more about the various types of functions in Excel, click the keys in the interactive below.

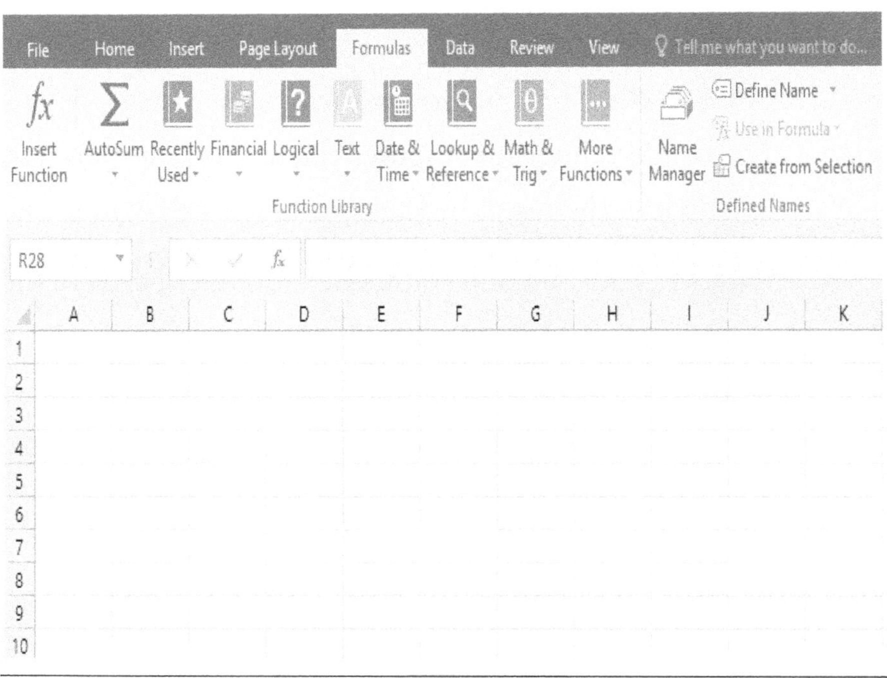

1.6 Some time-saving ways to insert data into Excel

There are five popular methods for adding simple Excel formulas while analyzing results. Each technique has distinct advantages. Therefore, before delving further into the primary formulations, this topic defines certain approaches so you can begin to create your preferred workflow.

1. Simple insertion: Composing a formula inside a cell

The simplest way to insert basic Excel formulas is to type them directly into the formula bar or cell. Typically, the procedure begins with the typing of an equal sign accompanied by the name of an Excel feature.

Excel is very intelligent in that it displays a pop-up function hint as you begin typing the function's name. This is the chart on which you can choose your preference. However, refrain from pressing the Enter key. Other than that, click the Tab key to begin inserting additional choices. Otherwise, you can encounter an invalid name mistake, which often appears as '#NAME?' Simply re-select the cell and navigate to the formula bar to complete your function.

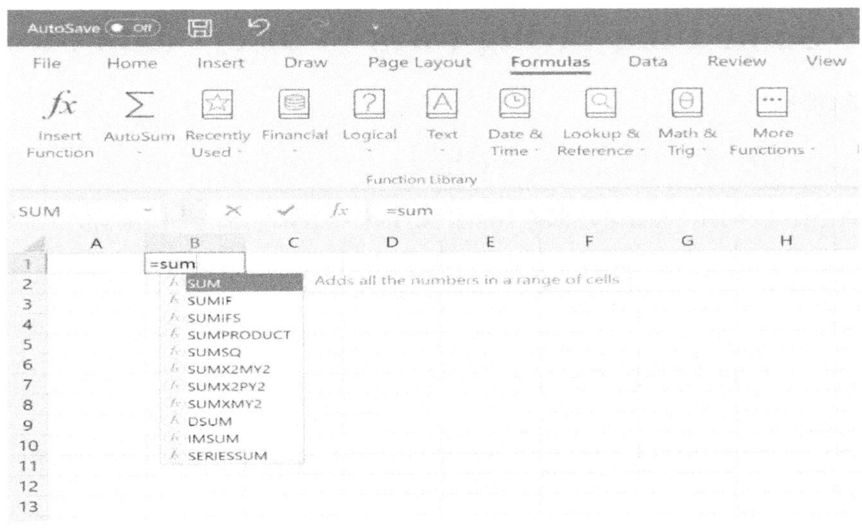

2. Choosing insert function from the formulas tab

If a person wants total control over the insertion of functions, the MS Excel insert function box is what you need. To do so, navigate to the formulas tab and click the first menu, Insert Function. Most of those functions necessary to complete the financial report will be included in the dialogue box.

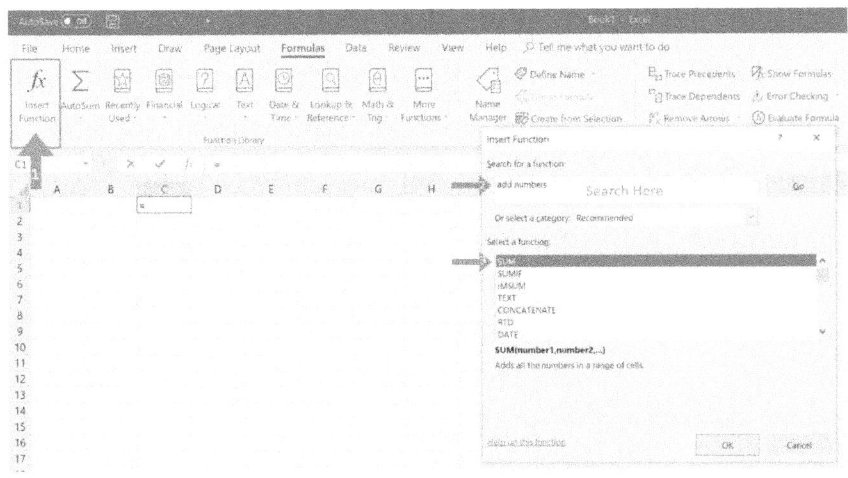

3. Selecting a formula from one of the groups in formula tab

This is the preferred method for those who want to access their favorite functions easily. To use this menu, move to the Formulas tab and click on the group of your choice. To show a sub-menu containing a list of functions, click. You can then pick your preference. If, however, your preferred category is not mentioned on the page, press the More Functions option, it is certainly hidden there.

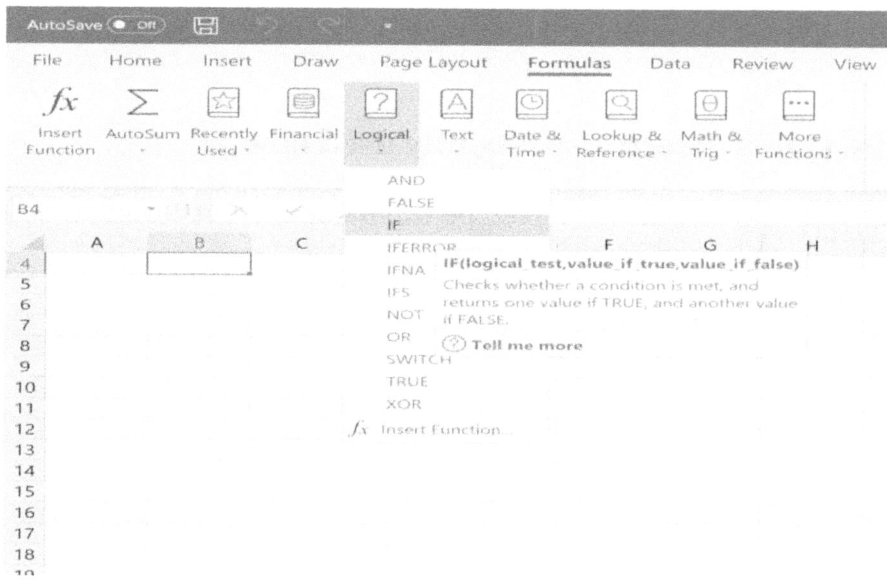

4. Using Autosum option

AutoSum is your go-to function for fast and routine tasks. Therefore, navigate to the Home tab and press the AutoSum option in the far-right corner. Then press the caret to reveal any additional secret formulas. Additionally, this feature is available in the Formulas tab, immediately following the Insert Function option.

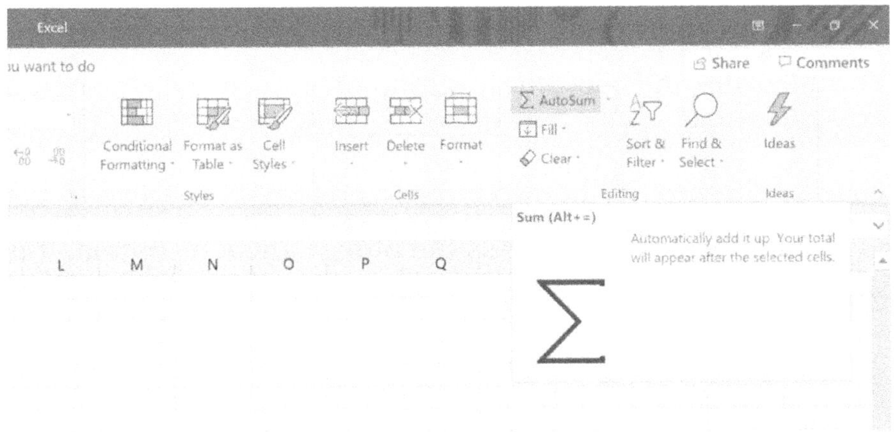

5. Quick Insert: Use recently used tabs

If you find it tedious to re-type the most current formula, you can use the Recently Used menu. It is a third menu option next to AutoSum on the Formulas tab.

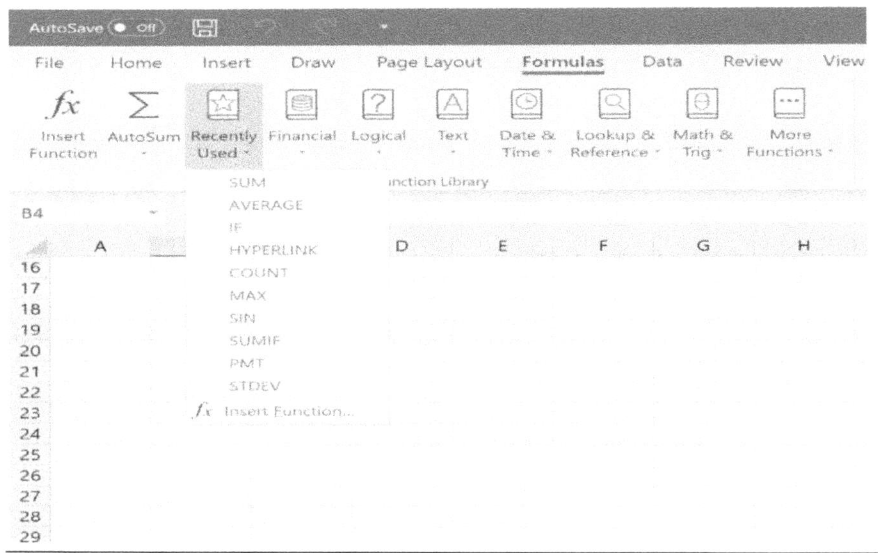

1.7 Microsoft Excel shortcut keys

Even if you are already acquainted with Microsoft Excel, you can be shocked at the amount and range of keyboard shortcuts available to help you function faster and more conveniently. Is anybody expecting you to memorize any of these keyboard combinations now? Certainly not! Since everyone's needs are unique, some would be more beneficial to you than others. And even though you just learn a few different techniques, the effort is worthwhile.

Additionally, this book has attempted to keep the list simple and clear. The following is a comprehensive collection of Excel keyboard shortcuts that can help you perform quicker. You may either scroll down the collection or use the Index to find the section you are looking for easily.

1. **General Program shortcuts:**

First, let's review some popular keyboard shortcuts for navigating workbooks, accessing help, and performing a few other interface-related tasks.

- **Ctrl+N**: Create a new workbook
- **Ctrl+S:** Save a workbook
- **Ctrl+O:** Open an existing workbook
- **F12:** Open the Save As dialog box
- **Ctrl+F4:** it close Excel

- **Ctrl+W:** Close a workbook
- **F4:** It repeats the last action or command. For example, if a last thing a person typed in a cell is "hi," or if a person changes the color of fonts, by clicking another cell and then clicking F4 repeats the action in the new cell.
- **Ctrl+Z:** Undo an action
- **Shift+F11:** Insert a new worksheet
- **Ctrl+Y:** Redo an action
- **F1:** Open the Help pane
- **Ctrl+F2:** Switch to Print Preview
- **Alt+Q:** Go to the "Tell me what you want to do" box
- **F7:** Check spelling
- **Shift+F9:** Calculate active worksheets
- **F9:** Calculate all worksheets in all open workbooks
- **Alt or F10:** Turnkey tips on or off
- **Ctrl+Shift+U:** Expand or collapse the formula bar
- **Ctrl+F1:** Show or hide the ribbon
- **Ctrl+F9:** Minimize the workbook window
- **Alt+F1:** Create an embedded bar chart based on select data (same sheet)

- **F11**: Create a bar chart based on selected data (on a separate sheet)
- **Ctrl+F:** Search in a spreadsheet, or use Find and Replace
- **Alt+H:** Go to the Home tab
- **Alt+F:** Open the File tab menu
- **Alt+N:** Open the Insert tab
- **Alt+M:** Go to the Formulas tab
- **Alt+P:** Go to the Page Layout tab
- **Alt+A:** Go to the Data tab
- **Alt+W:** Go to the View tab
- **Alt+R:** Go to the Review tab
- **Alt+X:** Go to the Add-ins tab
- **Ctrl+Tab:** Switch between open workbooks
- **Alt+F11:** Open the MS visual basic for the applications editor
- **Alt+Y:** Go to the Help tab
- **Alt+F8:** Run, create, delete or edit a macro
- **Shift+F3:** Insert a function

2. **Moving around in a worksheet or cell:**

You may use the keyboard shortcuts to move effortlessly through a worksheet, a cell, or an entire workbook.

- **Up/Down Arrow:** Move one cell up or down
- **Left/Right Arrow:** Move one cell to the left or right
- **Ctrl+Left/Right Arrow:** Move to the farthest cell left or right in a row
- **Ctrl+Up/Down Arrow:** Move to the top or bottom cell in a column
- **Shift+Tab:** Go to the previous cell
- **Tab:** Go to the next cell
- **Ctrl+End:** Go to the most bottom right used cell
- **Home:** Go to the leftmost cell in the current row or go to the start of the cell if editing a cell
- **F5:** Go to any cell by pressing F5 and typing the cell name or cell coordinate.
- **Ctrl+Home:** Move to the start of a worksheet
- **Alt+Page Up/Down:** Move one screen to the right or left in a worksheet
- **Page Up/Down:** Move one screen up or down in the worksheet

- **Ctrl+Page Up/Down:** Move to the previous or the next worksheet

3. **Selecting Cells**

As mentioned previously, you use the arrow keys to navigate between cells and the Ctrl key to change your movement. By modifying the arrow keys with the Shift key, you may extend the selected cells. Additionally, there are a few additional combinations for speeding up your selection.

- **Shift+Space:** Select the entire row
- **Shift+Left/Right Arrow:** Extend the cell selection to the right or left
- **Ctrl+Shift+Space:** Select the entire worksheet
- **Ctrl+Space:** Select the entire column

4. **Editing Cells**

Excel also provides keyboard shortcuts for the editing cells.

- **F2:** Edit a cell
- **Ctrl+X:** Cut contents of a cell, selected cell range or selected data
- **Shift+F2:** Add or edit a cell comment
- **Ctrl+C or Ctrl+Insert:** Copy contents of a cell, selected cell range or selected data
- **Ctrl+Alt+V:** Open the Paste Special dialog box

- **Ctrl+V or Shift+Insert:** Paste contents of a cell, selected data, or selected cell range
- **Delete:** Remove the contents of a cell, selected cell range or selected data
- **F3:** Paste a cell name if the cells are named in the worksheet
- **Alt+Enter:** Insert a hard return in a cell while editing a cell
- **Esc:** Cancel an entry in a cell or the formula bar
- **Alt+H+D+C:** Delete column
- **Enter:** Complete an entry in the formula bar or a cell

5. Formatting Cells

Are you ready to format any cells? These keyboard shortcuts simplify the method!

- **Ctrl+B:** Add or remove bold to the contents of a cell, selected cell range or selected data
- **Ctrl+U:** Add or remove underline to the contents of the cell, selected cell range or selected data
- **Ctrl+I:** Add or remove italics to the contents of the cell, selected cell range or selected data
- **Alt+H+H:** Select a fill color
- **Ctrl+Shift+&:** Apply outline border

- **Alt+H+B:** Add a border
- **Ctrl+Shift+_ (Underline):** Remove outline border
- **Ctrl+0:** Hide the selected columns
- **Ctrl+9:** Hide the selected rows
- **Ctrl+Shift+%:** Apply percent format
- **Ctrl+5:** Apply or remove strikethrough
- **Ctrl+1:** Open the Format Cells dialog box
- **Ctrl+Shift+$:** Apply the currency format

1.8 Some important Excel formulas

Excel formulas allow you to discover relationships between values in your spreadsheet's cells, conduct mathematical calculations on those values, and return the result in the cell of your preference. Sum, division, percentage, subtraction, average, and even dates/times will be automatically performed. Let's look at it in detail.

1. SUM

Each Excel formula begins with the equals symbol, =, accompanied by a text tag specifying the formula you want Excel to execute. The SUM formula in Excel is one of the simplest formulas you may use to calculate the sum (or total) of two or more numbers. To use the SUM formula, insert the values to be added in the format =SUM (value 1, value 2, etc.). The SUM formula accepts values that are either real numbers or equivalent to the value in a certain cell of your spreadsheet.

2. IF

In Excel, the IF formula is denoted by the formula =IF(logical test, value if true, value if false). This enables you to insert text into a cell "whether" another column in your spreadsheet is true or incorrect. For instance, =IF(D2="Gryffindor","10","0") will give cell D2 ten points if it contained the term "Gryffindor."

Value_ if_True: If the value is true – that is, if the student is a supporter of Gryffindor – this is the value that should be shown. In this scenario, we want it to be a ten to show that the student received ten marks. Use quotation marks only if you want the output to be text rather than a number.

Value_ if _False: If the student doesn't really live in Gryffindor, we want the cell to show "0," which equals zero points.

3. Percentage

To use the percentage calculation in Excel, enter the cells for which you want to calculate the percentage in the format =A1/B1. To convert the corresponding decimal value to a percentage, select the cell, click the Home tab, and then click the "Percentage" from the numbers dropdown.

Although there is no specific Excel "formula" for percentages, Excel simplifies the amount of any cell to a percentage, saving you from having to calculate and reenter the numbers manually.

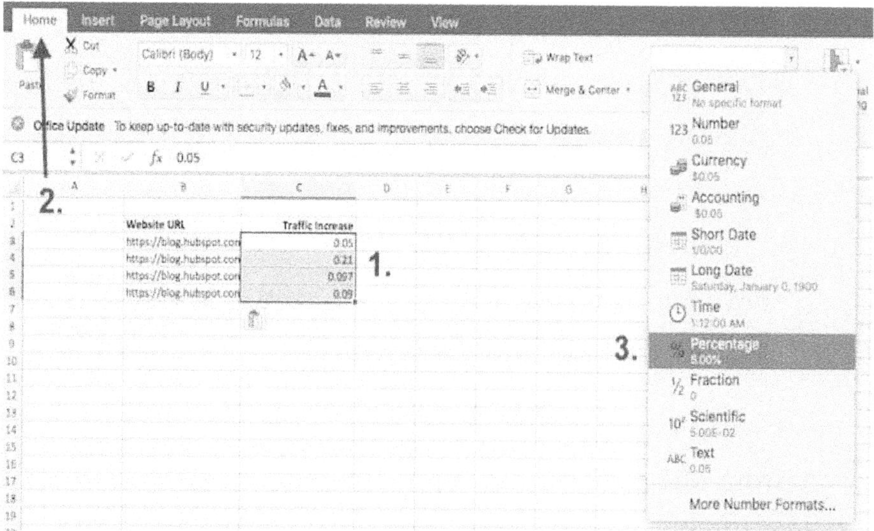

4. Subtraction

To use the Excel subtraction algorithm, insert the cells to be subtracted in the format = SUM (A1, -B1). This would deduct a cell using the SUM formula by preceding the cell to be subtracted with a negative symbol. For instance, if A1 was ten and B1 was six, the formula would perform ten plus six, returning four.

As for percentages, subtracting in Excel does not have its algorithm, but it does not imply it cannot be achieved. You may deduct some value (or those contained inside cells) in either of two forms.

C2 fx =SUM(B2, -A2)

	A	B	C	D
1	Value 1	Value 2	Result	
2	75	85	10	
3				
4				
5				

- **Making use of the = SUM formula:** To deduct different values from each other, insert them in the format = SUM (A1, -B1), with a negative sign (denoted by a hyphen) preceding the cell whose value is being subtracted. Press Enter to retrieve the difference between the two parenthesized cells.
- **Using the format: =A1-B1:** To deduct several values from one another, try typing an equals sign followed by the first value or cell, a hyphen, and a value or cell from which you want to subtract. To obtain the difference between the two values, press Enter.

5. Multiplication

To multiply cells in Excel using the multiplication formula, enter the cells to be multiplied in the format =A1*B1. This formula multiplies cell A1 by cell B1 using an asterisk. If A1 was 10 and B1 was 6, for example, =A1*B1 would return a value of 60.

You may believe that multiplying values in Excel requires a separate formula or that the "x" character is used to denote the multiplication of multiple values. Indeed, it is as simple as an asterisk — *.

SUM X ✓ fx =A2*B2*C2

	A	B	C	D	E
1	Value 1	Value 2	Value 3	Result	
2	75	85	95	=A2*B2*C2	
3					
4					
5					

In an Excel spreadsheet, highlight the empty cell to multiply two or more values. Then, in the format =A1*B1*C1..., insert the values or the cells you want to multiply together. The asterisk multiplies each value in the formula.

6. Division

To divide cells in Excel using the division algorithm, insert the cells to be divided in the format =A1/B1. This formula divides cell A1 by cell B1 using a forward slash, "/." If A1 was 5 and B1 was 10, for example, =A1/B1 will yield a decimal value of 0.5.

The division is one of the easiest features of Excel. To do so, highlight an empty cell and type an equals symbol, "=," followed by the two (or more) values you want to partition, separated by a forward slash, "/." The outcome should be in the following format: =B2/A2, as seen in the picture below.

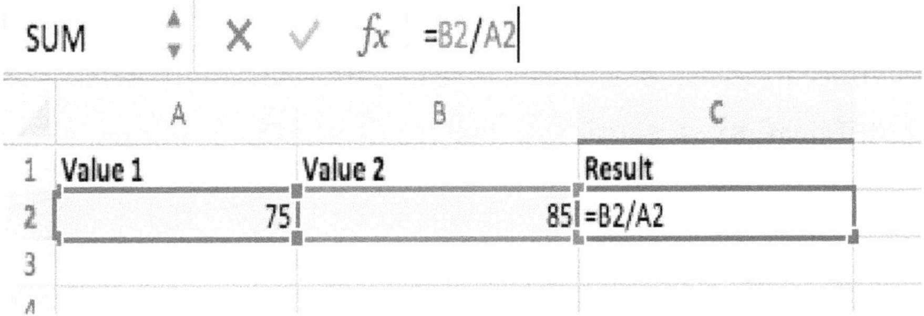

When you press Enter, the target quotient will appear in the cell you highlighted initially.

7. VLOOKUP

This is an oldie, but goodie and it goes into more detail than any of the other formulas on this topic. However, it is particularly useful because you have two sets of data in two separate spreadsheets and need to merge them into a single spreadsheet. The formula for VLOOKUP is **VLOOKUP (lookup value, table array, column number, [range lookup])**

Note: By following this formula, you must ensure that at least one column of both spreadsheets is identical. Analyze the data sets to ensure that the column of data you are using to merge the data is identical, with no additional spaces.

8. Date

The DATE formula in Excel is denoted by "=DATE (year, month, day). This formula will return a date that matches the values specified in the parentheses — including values

Alluded to from other cells. For instance, if A1 equals 2018, B1 equals 7, and C1 equals 11, =DATE (A1,B1,C1) equals 7/11/2018.

Occasionally, creating dates in cells of an Excel spreadsheet may be a difficult job. Fortunately, there is a convenient algorithm that simplifies the process of formatting dates. This formula can be used in two ways:

- Days may be generated using a sequence of cell values. To do so, highlight an empty cell, type "=DATE," and then insert the cells whose values produce your desired date in parentheses — beginning with the year, followed by the month number, and finally the day. The final format will be =DATE (year, month, day).
- You can automatically set today's date. To do so, select an empty cell and insert the following text string: =DATE(YEAR(TODAY()), MONTH(TODAY()), DAY(TODAY()). By pressing enter, the latest date in your Excel spreadsheet would be returned

D4 fx =DATE(A4, B4, C4)

	A	B	C	D	E
1	Year	Month	Day	Date	
2	2018	12	2	12/2/18	
3	2018	2	18	2/18/18	
4	2018	7	11	7/11/18	
5					
6					
7					

1.9 Mistakes to avoid while dealing with formulas in Excel:

1. Recall the laws for addition, multiplication, division, and subtraction using brackets (BODMAS). This means that the statements enclosed in brackets are evaluated first. In arithmetic operators, the division is verified first, followed by multiplication, addition, and eventually subtraction. Using this rule, we can rewrite the previous formula as =(A2 * D2) / 2. This means that A2 and D2 are evaluated first, followed by a two-step separation.

2. Usually, Excel database calculations use numeric data; you may use data validation to specify the kind of data that a cell can recognize, such as just numbers.

3. On the screen, press F2 to verify that you are working with the correct cell addresses in the formulas. This will highlight the formula's cell addresses, helping you to double-check if they are right.

4. When working with many rows, you can give each row a unique serial number and keep track of the totals at the bottom of the sheet. To ensure that the formulas included all rows, compare the serial number count to the total number of records.

1.10 Excel tricks that can turn someone into an Excel expert

1. **Eliminate Duplicates from a Range of Cells**

   Remove redundant objects from vast data sets to guarantee that only one piece of data exists. Click on any row and column that can include duplicate data sets to delete them. After selecting a column or row, press the Data tab, then choose Remove duplicates from the Data Tools.

2. **Convert the rows into columns**

   Transpose the data set from rows to the columns with ease. To do so, highlight the range you would like to modify, right-click it, and select copy. After copying the selection, navigate to the location where you want the cells to go and right-click the cell. Select Paste Special from the right-click option, and then verify Transpose at the bottom of your window.

3. **Alternate between the two opened Excel files**

   Click on CTRL+Tab which allows you to easily move between the two open Excel files.

4. **Alphabetize data in a row or column**

   Sort data in the columns and rows alphabetically by highlighting your selection, right-clicking it, and then selecting Sort. Pick Sort A to Z or Sort Z to A once the window opens.

5. **Freeze the cells**

by freezing the contents of cells, you can lock them in place when scrolling. Simply click on the View tab and then click Freeze Cells. You may freeze a custom set of cells or only the first column or top row.

6. **Wrap the text to fit**

Wrap Text prevents long texts from spreading into adjacent cells. To wrap text, pick the Home tab and then the Wrap Text option from the drop-down options.

7. **Use Excel's autofill feature**

if you input data into cells in a consistent pattern (such as serial numbers or dates), Excel will automatically fill the information, saving you time and effort. To do so, simply pick a set of data and wait for a + symbol to appear in the bottom right corner. When it does, drag the mouse in the direction you want the autofill to go.

8. **Create Macros to save time on repetitive processes**

Create useful macros to simplify time-consuming tasks and save you time and effort. To do so, select the Developer tab and then Macro Recording. Continue performing any tasks you want to automate, then click Stop Recording to save the macro.

9. **Start a new line in the cell**

simply press Alt+Enter to begin a new line in a cell.

10. **Add multiple rows or columns**

select the rows or columns into which new cells should be inserted, then right-click on the list. Insert new blank cells between your previous data sets by clicking insert.

11. **Utilize keyboard shortcuts**

Utilize time and effort-saving keyboard shortcuts. Copy with CTRL+C, paste with CTRL+V, and pick all data points with CTRL+A.

12. **Find and replace data**

every piece of data can be replaced with a specified value. To do so, press the Home tab, and then scroll to the toolbar's right and then Find & Select. Once the drop-down menu emerges, press Replace and specify the words you want to search and replace in the given boxes.

13. **Expand the formula bar**

Expand your formula bar to see formulas or entered content more clearly. To expand the bar's size, simply press the downward pointing arrow on far right side of the bar.

14. **Hide the specified rows**

to hide a certain row for improved viewing, simply pick it and right-click on it. Select hide once the drop-down menu emerges.

15. **Collapse the toolbar ribbon**

to collapse the toolbar ribbon and increase the amount of room available for viewing cells, click CTRL+F1. This choice would then expose additional cells by removing the ribbon. Pick either of the tabs to re-display the toolbar ribbon, and then click the thumbtack present in bottom right to re-pin it to the screen.

16. **Encrypt and protect files**

Protect and encrypt Excel sheets to keep every amount of data confidential or personal. To do so, press the Review tab and then on the far right side of the toolbar, click Protect Sheet. Select which sections you want to secure, then enter a password and click OK.

17. **Split the data into different cells**

Separate several data metrics from the single column to facilitate analysis and identification of key components. To divide the data, press the Data tab and then choose Text to Columns. Once the box appears, you can specify the location of the important separating points.

18. **Select an entire row**

simply click on a cell in the row and then press Shift Spacebar to pick the whole row of data.

19. **Share a common format between the cells**

With the Format Painter function, you can convert the format of a cell to that of another. To copy the format, right-click on any cell that contains the desired format and then select the Home tab. Once on the Home page, press Format Painter and pick the additional cells whose formats you want to modify.

20. **Add headers and footers**

when printing documents, use headers and footers to improve worksheet clarity. To add a header, press the Insert tab and then choose the Header and Footer from the Toolbar's Text section.

Chapter No: 2 Versions of Microsoft Excel

If asked what spreadsheet application you use, you are likely to say Microsoft Excel. However, computers have advanced significantly since Excel's initial release in 1985. The specialist experts are familiar with the various features of the various models and are capable of developing bespoke solutions for all of them. Since 1985, approximately 29 variants of Excel have been published, from 1985 to the present. The majority of persons today have only Excel 2016, Office 365 (2021) or 2019. Although Excel might not be around in perpetuity, it is truly impossible to imagine a world without it. And it undoubtedly altered the course of history during its reign.

Microsoft took a lesson from the first time Lotus demolished Multiplan in the sector. It set them on the road to creating a game-changing app package that adds value to companies worldwide. It put them on the road to developing a world-class operating system in which to integrate it. Microsoft capitalized on its early corporate errors and fundamentally altered the way our civilization does business.

Almost every Excel version differed substantially from the previous one, just as MS Windows operating systems and Mac are today. The following is a list of older and recent additions to Microsoft Excel.

2.1 Microsoft Excel 1985

On September 30, 1985, Microsoft launched the first edition of Excel for the Macintosh, and in November 1987, Microsoft released the first version of Excel for Windows, version 2.05 (to synchronize with Macintosh version 2.2). Microsoft Excel is a spreadsheet program developed by the Microsoft Corporation in 1985. Excel is a widely used database program that organizes data in columns and rows and can be modified using formulas to execute mathematical operations on the data. The influential software gained popularity due to its graphics-heavy interface, designed to operate on the new Windows computers. Due to Lotus's delay in releasing a Windows implementation of its database, Excel gained market share and gradually became the leading spreadsheet technology in the mid-1990s. Later iterations of Excel saw major enhancements such as outlining, toolbars, painting, three-dimensional charts, various shortcuts, and increased automation. In 1995, Microsoft modified Excel's naming convention to emphasize the product's initial year of release. The suite's initial version of Microsoft included Excel 2.0, Word 1.1, and PowerPoint 2.0. According to the magazine Info World's October 1, 1990 report, "all three applications support Dynamic Data Exchange (DDE), which enables programs to use Windows services to exchange data in real-time." This ensures that these three programs will communicate with one another, bringing a digital office into existence.

2.2 Microsoft Excel 2002 and Microsoft Excel 2003

Nowadays, relatively few companies choose to use Excel 2000 or Excel 97. Excel 2002 is the original XP edition of Excel, which is somewhat close to the current versions in terms of features. Prior to Windows XP, Excel could not be completely implemented into the Windows experience; for example, it could not ask the user for a file location with the standard Windows file explorer.

Generally, Excel 2002 and 2003 are similar. Microsoft did provide a few additional features in 2003, such as "WORKDAY," although they were only available in subsequent versions. If, though, you are attempting to generate spreadsheets for Excel 2003 in a later edition of Excel, you can save files in the Xls format, not the newer defaults, such as xlsx or xlsm. The latest file types can be accessed in earlier versions only after installing a compatibility kit. Fortunately, the older Xls format has few disadvantages, except that the files are slightly larger.

Additionally, avoiding excessive color when creating files for earlier versions of Excel is suggested. Excel has a quite small color palette before Office 2007, and although it can try to color match, the effects will occasionally be very ugly.

2.3 Windows Microsoft Excel 2007

Excel 2007 was also only usable on Windows-based PCs. Excel 2007 was a major upgrade from previous releases, introducing the Ribbon interface and converting the file format from the standard.xls to the more modern.xlsm and.xlsx formats. Excel files can now contain approximately 1 million rows (up from the previous limit of 16,384), and security has been significantly enhanced. Excel's charting capabilities have also been significantly improved in this update. According to polls, many users initially disliked the new interface, but Microsoft stuck with it, and most users still do not choose to switch.

2.4 Mac Microsoft Excel 2008

Excel 2008 was also available exclusively on Mac Apple computers. However, appearances may be deceiving—Excel 2008 is a complete rewrite built to operate natively on both Intel-powered Macs and PowerPC-based machines. Additionally, this version of Excel has several additional capabilities, but it eliminates at least one significant area of functionality. As a consequence, Excel 2008 could leave you unimpressed.

2.5 Microsoft Excel 2010

Current versions of Excel are very distinct from older versions. Rather than utilizing the standard menu bar, you can now access functions through a graphical ribbon on the top of the page. Despite this, the functionality is strikingly identical to previous models. Maybe the most significant enhancement is the addition of support for spreadsheets of more than 65536 rows and 256 columns. Today, the computational capacity exists to evaluate a database of 100,000 names, and Excel has developed to the point that it can solve this type of problem. Some significant updates include the default file types and paint palette. Another default setting that has been altered is the default font. Calibri was introduced as the default font in Office 2007 in place of Arial. It appears to display smaller, which is advantageous when reading through several columns. Additionally, it has a more modern sound.

It is worth mentioning that certain features, such as Macro security settings and show/hide sheet tabs are hidden from the ribbon, making them less visible than previous models. Excel 2007 and 2010 have a lot in common. The significant improvements in 2010 are more evolutionary, as Microsoft works to transition the whole Excel program online, most notably through SharePoint.

2.6 Mac Microsoft Excel 2011

This was the last Mac edition of Excel to differ from the Windows version in terms of name. Excel 2011 was released only on Mac computers. This, like Excel 2010 and Windows, was the 13th edition of Mac Excel; however, version 13 was omitted for superstitious reasons, while Excel 2011 was revealed to be the 14th version.

2.7 Windows Microsoft Excel 2013

Excel 2013 introduced new Slicers, 50 new features, and the Flash Fill feature, but it was limited to Windows computers. Excel 2013 is a robust spreadsheet program that is ideal for practical usage in a variety of industries. A spreadsheet may be used to keep track of lists or to create complex reports. The true strength of it comes from the ability to interpret, modify, and quantify data in a spreadsheet dynamically.

2.8 Microsoft Excel 2016 and Microsoft 365 for windows

Perhaps for the first time before 2000, Microsoft decided to give the Mac and Windows variants of Excel the same name. Even so, these two programs are very distinct, in part due to the iOS operating system's unique user interface. Rather than releasing new functionality for each new edition, Microsoft began distributing daily feature updates over the internet for Excel 2016. Users with bundled retail models of Excel now have a somewhat different Excel version than Office 365 since these new enhancements were formerly open exclusively to Office 365 customers. Excel 365, like Windows 11, is designed to have an eternal existence. Rather than the previous three-year upgrade period, 365 is a rapidly changing software with new improvements included with each new edition.

Because enterprise customers dislike continuous change (and need rock-solid, stable releases), Microsoft launched the idea of "update channels" for Office 365. Business customers have the choice of subscribing to a bimonthly upgrade channel. This resulted in a reliable and tested "Semi-annual" version released in January and July of each year. The Smart Method accommodates Office 365's ever-changing existence by releasing new editions of our books every six months to coincide with each new update. Microsoft had recently released the fourth version (covering the July 2020 semi-annual update). After 2019, Excel 365 has introduced some incredible functionality, but nothing compares to the inclusion of Dynamic Array functions and Dynamic Arrays in the July 2020 version. This is not a minor adjustment. Excel's engine needed to be updated to support the current dynamic array concept.

In Excel 365, many existing Excel functions will now be used in completely different ways, although array-aware functions have substituted several long-standing favorites. For instance, VLOOKUP has been superseded by XLOOKUP (a modern array function that is easier and more versatile to use and more powerful). Excel 365's latest interactive array capability elevates it beyond Excel 2019. Dynamic arrays are inherently incompatible with Excel 2019 and will result in modern workbooks (created using Excel 365) failing to function properly in legacy models (in this context, Excel 2019 is a legacy version even though a person can still purchase it). Excel 365 is the latest, greatest and most powerful Excel version you can use, and it is available for a very modest monthly subscription.

Microsoft 365 Personal subscriptions include Word, PowerPoint, Excel, OneNote, Publisher, Outlook, and Access. They can be accessed on up to five computers through PC and Mac. Per month, subscribers obtain an extra 1 TB of One Drive storage and 60 free Skype minutes.

Up to six users can share subscriptions to Microsoft 365 for Family. It contains all of the functionality included with the Personal plan and extra One Drive storage (6 TB).

Microsoft 365 Business programs provide Office applications in addition to improved compliance protections and connections to Microsoft Teams.

Microsoft launched 365, intending to eliminate one-time sales gradually. If you do not refresh Office regularly, the conventional product should perform fine. However, if you are looking to get keys to unique upgrades, 365 is the way to go. Regardless, keep in your mind that Microsoft has stated that the standard Office Suite will be phased out in the not-too-distant future.

2.9 Microsoft Excel 2019

Excel 2019 has a slew of innovative and improved functionality and capabilities. Here are 12 of those enhancements, which you can personally download, install, and test: Excel contains the following six modern or modified functions: IFS, -TEXTJOIN, MAXIFS, SWITCH, CONCAT, and MINI FS are all available functions. SWITCH and IFS are simpler variations of the nested IF feature, obviating the need for more complex nested functions. Although both methods are true, the IFS approach needs only one function, while the IF approach requires three. The IFS approach results in a somewhat longer formula in this case, but the IFS approach allows you to set out each criterion, while the IF approach does not. Numerous analysts suggested that Excel 2019 will never exist.

Microsoft left everybody waiting until the final moments before revealing that a new "perpetual license" edition of Office will be offered alongside Excel 365. Microsoft has embraced the SaaS (information as a service) model, in which software is leased rather than purchased. They seem to have also recognized that certain purchasers seem to prefer the "buy once, use forever" model. Numerous analysts now speculate that Excel 2019 is the last perpetually licensed edition. Excel 2019 is now significantly behind Excel 365, with the "killer version" of complex arrays open exclusively to Excel 365 subscribers since no additional features will be introduced to Excel 2019.

The most significant change in Excel 2019 is the addition of the "power" tools Power Pivot, Power Maps (3-D Maps) in both models and Power Query (Get & Transform). These are extremely high-level OLAP tools that enable Excel to analyze "big data" and extend the capabilities of any Excel version to conduct "modern data analysis." Having said that, since Excel 2019 was launched, the "power" tools have expanded further, with additional "power" features accessible exclusively to Excel 365 users.

2.10 Microsoft Office 2021

As you are no longer conscious, Microsoft enhanced Excel 2019 with a slew of new features. Additionally, the tech giant continues to upgrade the venerable spreadsheet program through Office 365 updates. Of course, these latest features are useless until you know them and understand how to utilize them. Continue reading to hear about five of the most important new features in Excel and how to use them. Microsoft is introducing a few notable updates to Office 2021. Can the latest upgrades make a difference in terms of the overall experience?

The news of Office 2021's opening has garnered considerable interest. This is mainly because nearly everybody now uses at least one of the many Microsoft Office systems. Thus, what did they modify? Can the latest updates to the Office suite improve the overall experience with the suite's services, or will they make matters significantly more complicated? If you are curious about the changes that Microsoft Office 2021 would bring, just continue reading this book.

2.11 Microsoft Excel for MAC

Excel was first designed for Macintosh computers in the mid-1980s. Regrettably, Microsoft has now withdrawn funding for this rival operating system. Excel 2004 was basically a repackaged variant of Excel 97 and therefore has a host of usability problems with the then-current Windows version. Excel 2008 for Mac updated the user experience to match that of the Vista edition of the software, but Microsoft omitted all support for Visual Basic from their Office suite. This ensures that all Windows-developed scripts or types cannot be used on a Mac. As a result, the experts often decline to create spreadsheets for Macs. Without the ability to write macros, it becomes very challenging for an experienced user to progress beyond the level of a novice user with a spreadsheet. Earlier Mac models contain the following: (2001, 2000, 1998, 2005, 2004, 2003, 2002 and 2001)

It is a little-known reality that the first edition of Excel was only available for Mac, even though Microsoft has an even earlier spreadsheet product named Multiplan for MS-DOS and some other console-based operating systems. Excel's first Windows release was simply a port of the Mac's "Excel 2". Although Microsoft recently launched Excel 2019 for Mac and Excel 2019 for Windows under the same names, they did so previously for Excel 2000, which was released for Windows and Mac. If you are already using an older edition of Excel for Mac, it is time to update! The Excel 2019 for Mac courses will tell you all you need to know about using the latest edition of Excel.

2.12 Today's Microsoft Excel

To the present day, with the introduction of Excel365 and Excel 2019, Microsoft Excel is by far the most familiar, flexible, and commonly adopted enterprise technology on the planet, owing to its ability to adjust to almost every business method. When combined with other Microsoft Office software, such as Outlook, Word, and PowerPoint, there is nothing that this extremely powerful mix cannot accommodate. The Office Suite and Microsoft Excel provide almost unlimited applications. Given below are the top ten list of the most frequently used and strong built-in Excel features:

1. Model and interpret nearly every data set efficiently

2. Easily zero in on the right data points

3. Create data charts inside a single cell

4. Almost everywhere, you can use your spreadsheets.

5. Collaborate to link, share, and achieve more

6. Increase the interactivity and dynamic nature of Pivot Charts.

7. Make the data presentations more sophisticated

8. Simplify and expedite tasks

9. Increase the computing resources to create bigger, more complicated spreadsheets

10. Use Excel Services to publish and share data

Add to that the potential to optimize and simplify any process by using VBA, and you have a massive value-added BI (Business Intelligence) tool that is adaptable and innovative enough to address almost any business requirement.

Are you interested in using Microsoft Excel to develop market solutions? Excel assists businesses of all sizes, from household brands to mom-and-pop stores, and optimize the workflows using Microsoft Excel and other Microsoft programs.

2.13 Microsoft Excel's Future

How can we proceed from here? With the internet at the center of our lives and businesses, it is natural for the interests of the many to win. When Microsoft platforms begin to grow, staying current on new technology becomes a full-time task. Microsoft Excel will maintain its role as the leading framework for data analysis, chart and presentation creation, and integration with powerful platforms for virtual dashboards and business intelligence workflows.

Businesses are increasingly relying on cloud computing for data accessibility and collaboration. This is where we can see Microsoft Excel's future developing at a breakneck pace over the next few years, enabling multi-user access to vast data for research, monitoring, and significant improvements in productivity and production. With the inclusion of functionality such as Power BI (Power Pivot, Power View, Power Query, and Power Maps), researchers, monitoring experts, and business intelligence professionals make Excel more applicable for full-fledged business intelligence implementations. Microsoft's latest Power BI Desktop offers makes it simple to build efficient dashboards that can be shared through several endpoints such as iPads within your enterprise. Since the simplest way to import data into Power BI is via an Excel file, Excel remains the most significant component of the Power BI workflow.

Excel is one of the programs that businesses cannot function without. It is the one program in which you do not have to be technically proficient and sound as though you are. As time is passing, Excel can evolve and make it easy to connect with enterprise-level databases while still allowing for smooth integration of business intelligence tools. Numerous business intelligence software programs, such as Oracle Answers (which uses ivy files to enable you to load the reports and data directly into Excel), and also heavy data visualization software, such as Elkview, have agreed to allow you to export data to Excel, because it is, after all, what business users want.

Excel must continue to do what is incredibly fantastic to be the world's favorite analytics application. This includes the following:

1 Include a sandbox-like environment in which everyone can experiment with data and generate information.

2 Keep the Excel software offerings minimalistic (with Power BI, purchasing and using Excel has become a complicated process).

3 Provide users with competent and compatible Excel apps for phone devices and tablets.

4 Provide features that allow users to perform powerful data analytics processes with a single click (as against a complex set of formulas, manual steps and pivot tables that most analysts do now).

Microsoft Excel is the standard method of choice for displaying and analyzing data, with over 750 million users worldwide. And it is now in a spot to become the latest BI weapon of preference. Microsoft announced several forward-thinking built-in Business Intelligence (BI) capabilities in Excel 2016. This is enormous in the era of large data.

However, this announcement posed some concerns in the area of big data manipulation. Is Excel sufficiently powerful? Is it capable of manipulating the amount and variety of data that companies today contend with? Is there a future for Excel in Business Intelligence workflows? Of course, the response is yes. Custom solutions are expected in today's dynamic market climate to retain an advantage over competitors and maximize income. Consultancies specializing in Microsoft Excel are the leading experts on current and new technology. Having an expert consultant on retainer is critical for fully using the strength and efficiencies needed to succeed in the twenty-first century. With Microsoft Excel installed on virtually every device in modern enterprises, its data collection and manipulation capability, along with its relative ease of usage, render it a powerful market analytics platform that is now accessible to the masses. For developers, Excel has long been an excellent platform for developing business intelligence applications for customers, but with a significant technological effort. Microsoft has spent extensively in the last few years developing a range of

Add-ins, culminating in complete product convergence, to render business intelligence accessible to all Excel customers. Users can now bind to and control their data even more easily thanks to the power resources such as Power Pivot, Power Map, Power Query, and Power View. You can Analyze and explore data and gain more information. It is capable of displaying the information more graphically and engagingly. Excel and business intelligence seem to have a rather promising future.

Due to the low cost and high capability of Excel, Business Intelligence applications are now readily accessible to small and medium-sized companies (SME's), enabling them to obtain useful perspectives that were formerly only available to large organizations with the resources to implement those workflows and processes. A basic illustration of this is the opportunity to create Business Intelligence dashboards in Excel to visualize data such as revenue within a given geographic area over a specified period using graphs and tables. A dashboard of this kind enables SMEs to benefit from a real-time and customizable solution that can aggregate data from various sources and provide a centralized page of key performance indicators, essential business data, and analytics on a single screen.

Converting vast quantities of data into actionable insights for the advantage of small and medium-sized businesses. The future has already arrived; all that remains is for us to spread the message. With Excel 2010/2013/2016, we can use Excel resources such as Power Query to extract data from virtually any source, filter it, and clean it before submitting it to Power Pivot, where we can build relationships between the various data sources, mash them up in PivotTables, and display it in fantastic, interactive Excel Dashboards. And to refresh your dashboard, simply click the 'Refresh Everything' icon. Completed! Excel is a cost-effective self-service business intelligence solution for all.

Excel is here to remain viable and integrates well with the internet and business intelligence platforms. What counts most is that prospective customers of business intelligence systems both small and large businesses conduct analysis and weigh their options before selecting a business intelligence provider to provide the best solutions for the business. Assuming that this entails treating Excel as the universal asset toot, Excel's future BI workflows are stable. So, businessmen at Excel Supersite believe that this pattern will continue to evolve as more SME's begin to self-manage their business intelligence requirements and fully leverage the strength of the data at their fingertips.

Excel is establishing a foothold in the field of Big Data BI. This is excellent news for the millions of data scientists worldwide... Excel is common to them, easily available, and stronger than ever. Says, Michael Alexander

Chapter No: 3 How to determine the version of Excel you are already using?

You are aware that Microsoft Excel features vary between versions and that there are several Excel versions, including Excel 2003, 2010, 2007, 2016, 2013 and many more! Thus, if you use Microsoft Excel, you must verify the version of the spreadsheet that you are using. In certain instances, you may need to recognize the version of Excel that you are already using. Instructions for some of the tips featured in Excel Tips differ according to the edition of Excel being used. If you are a complete beginner to Excel, you might be unsure how to assess the edition you are already using. This chapter will demonstrate how to find the Excel version information.

3.1 Find the version of Microsoft Excel 2007

1. Click the **Office** button and then **Excel Options**.

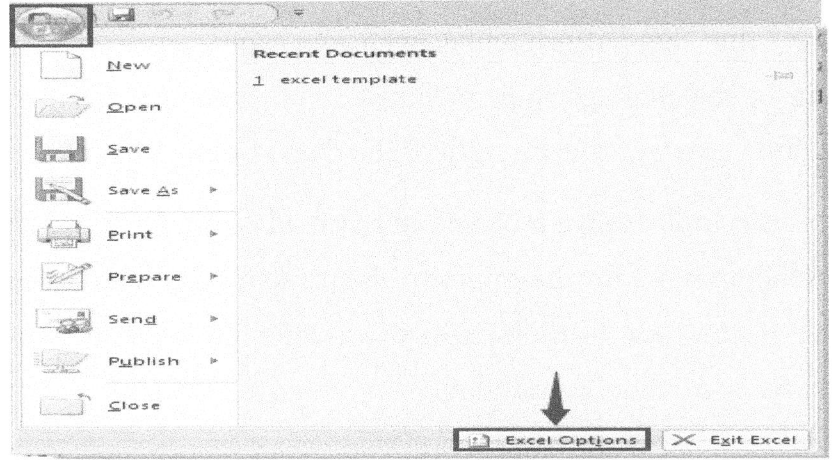

2. In the Excel Options dialogue box, press Resources in the left pane; then, in the about Microsoft Office Excel 2007 portion, you will find the Excel version. See the picture below:

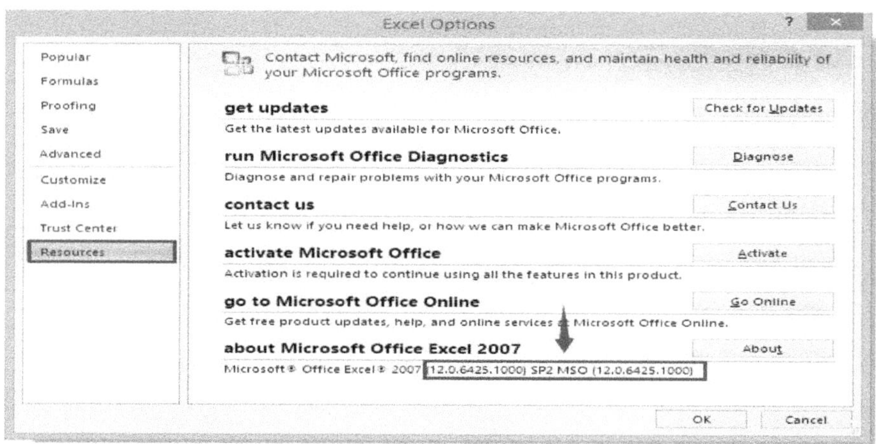

3.2 Find the version of Microsoft Excel 2010

For Microsoft Excel 2010, follow these steps:

1. Click on **File** and then **Help**. Then, in the About Microsoft Excel section, you will see the Excel version. See the picture given below:

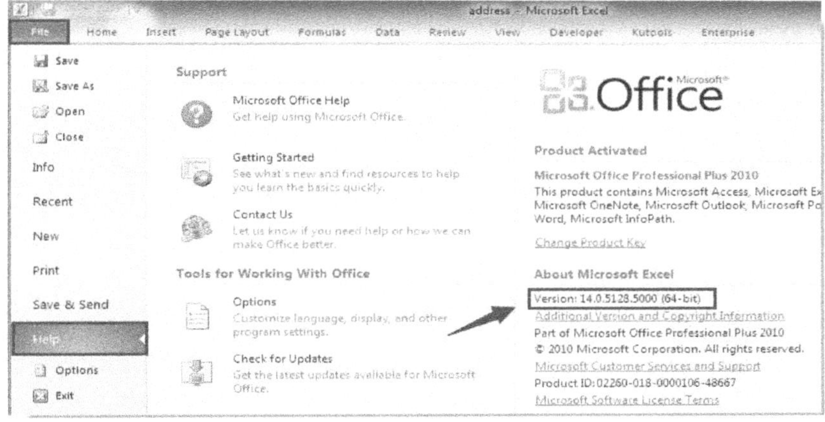

3.3 Find the version of Microsoft Excel 2013

To determine the version of Excel 2013, do the following steps.

1. Click the **File** then **Account** and then **about Excel**. See the picture below:

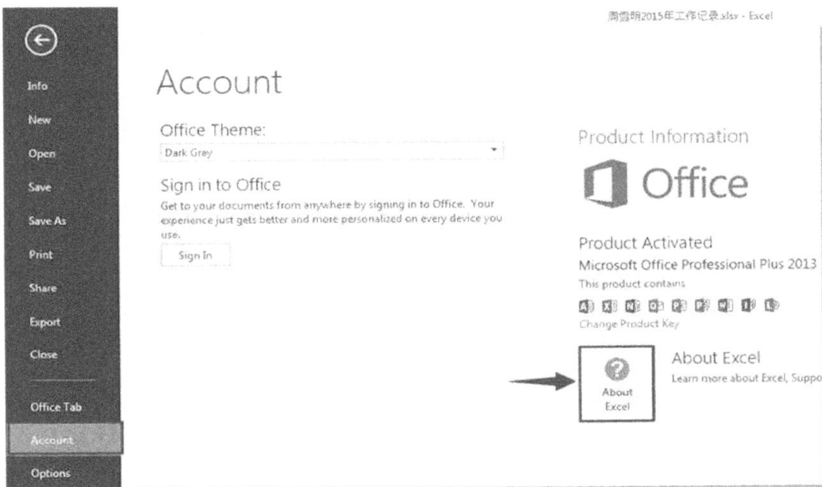

2. Then, in the About Microsoft Excel dialogue box, you will find the Excel version.

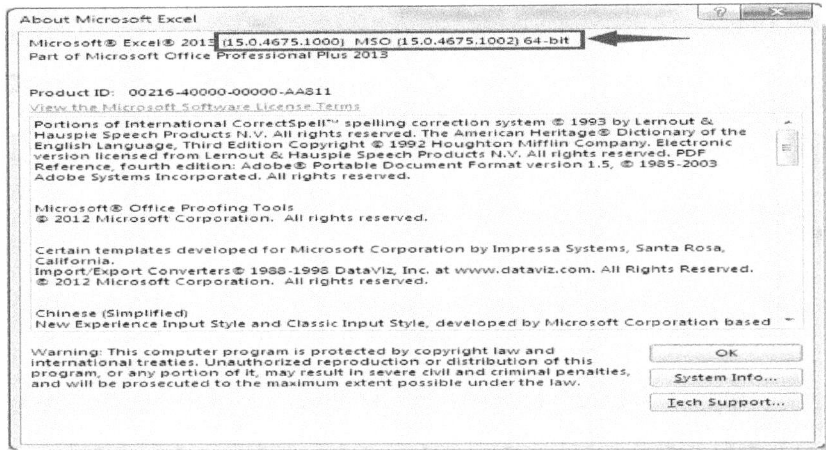

3.4 Find the version of Microsoft Excel 2016

For MS Excel 2016, the distinction is readily apparent. At the top, you can see the text "Tell me what you want to do. “This is the only edition of Excel that has this function. This text is not included in any other version.

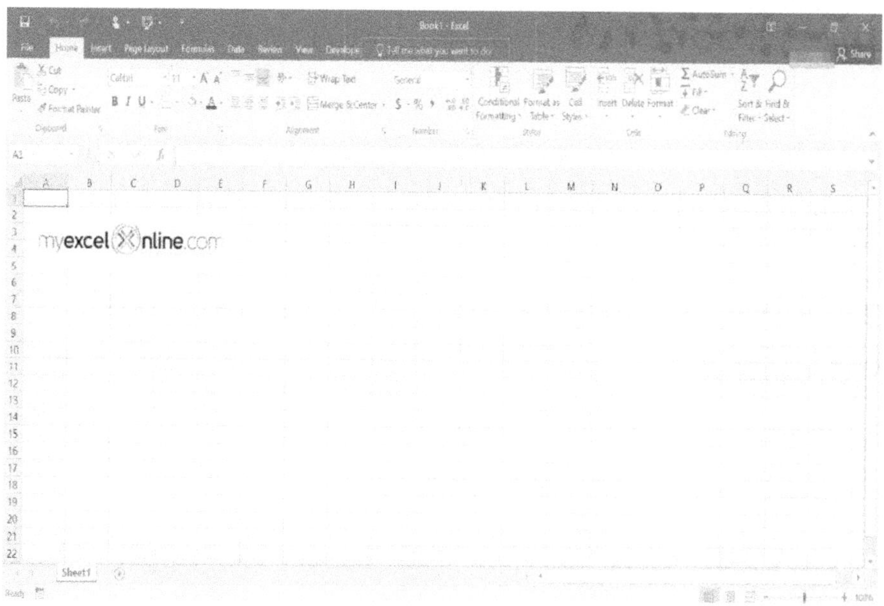

3.5 Find the version of Microsoft Excel for Mac

It is very easy to collect details about and verify the Excel Version on a Mac. To obtain details about the build number, product version, and so on, you may either look at the application's start screen, the menu ribbon, or the help or about option in Excel.

1. Click on the **Excel** on the Mac menu bar and then select **About Microsoft Excel**. See the picture below:

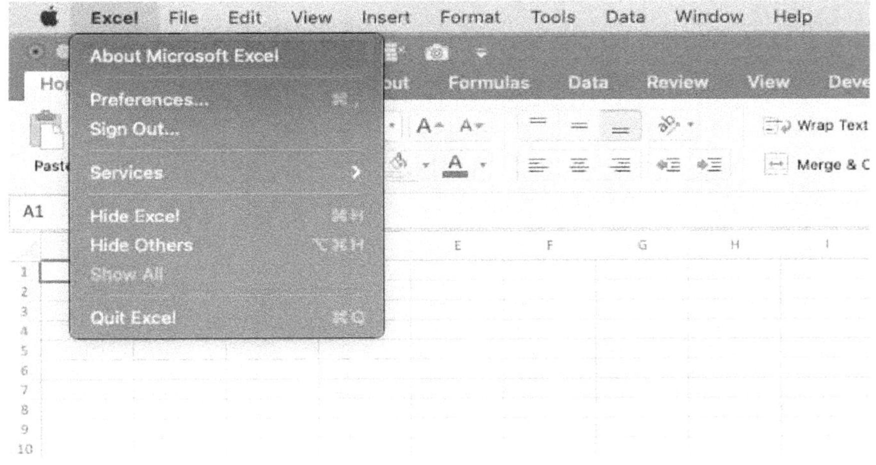

2. All info will be shown in the About Excel dialogue box!

3.6 Find the Microsoft Excel version with the VBA code

Apart from the methods described previously, you will even find an Excel version with a VBA code.

1. Click **Alt +F11** to open **MS basic for applications window**.

2. Click **Insert** and then **Module**. See the picture below:

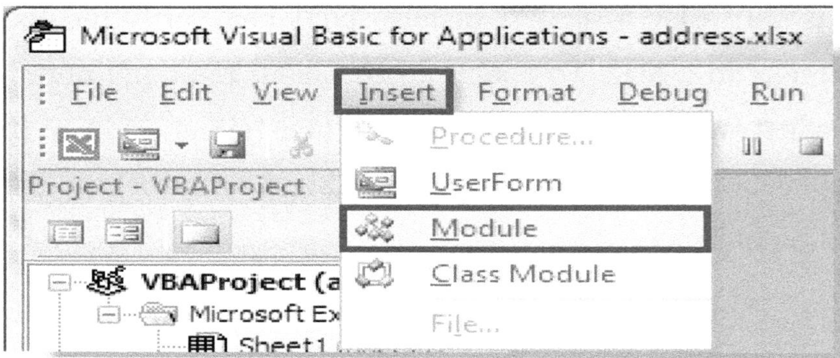

3. Copy and then paste the below VBA code in the Module window and then click the **F5** key to run this code.

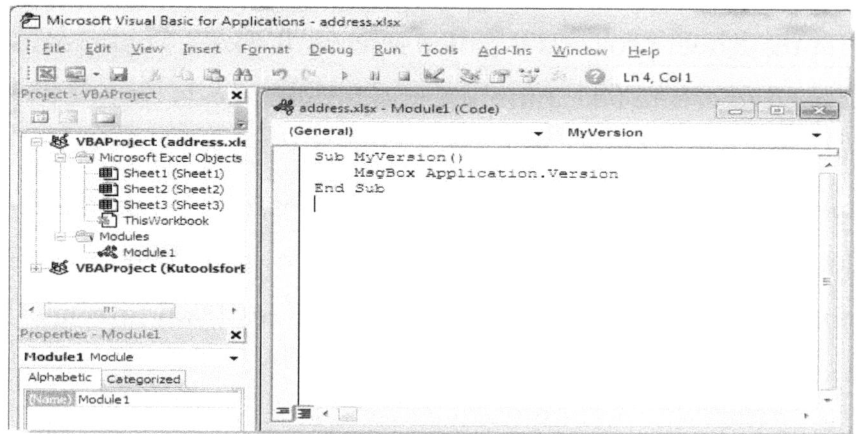

VBA code: find MS Excel version

MsgBox Application.Version

Sub MyVersion

End Sub

4. After this a **Microsoft Excel** dialog box will appear with the version number displaying.

Chapter No: 4 How to Use Microsoft Excel?

A spreadsheet program is called Microsoft Excel. That means it is used to create text, number, and formula grids that describe calculations. Many companies rely on it to keep track of spending and revenue, schedule schedules, chart statistics, and display financial statements concisely. However, you must first learn how to use Excel. This chapter will lead you through the process.

4.1 Use of the Ribbon

The ribbon in Excel provides shortcuts to commands. A command is a user-initiated action. Commands include printing a document, making a new document, and so forth. The Excel ribbon can be seen in the image below.

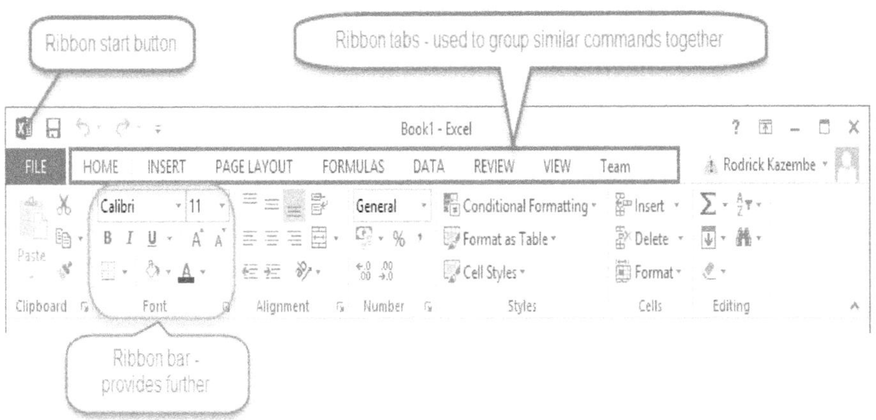

4.2 Components of the Ribbon

The ribbon is a Microsoft-created user interface feature that debuted with Microsoft Office 2007. The Home, Page Insert, References, Layout, and other tabs on the Microsoft Word ribbon, for example, each show a different set of commands when chosen. The ribbon's main components are as follows:

1. **Ribbon tabs**: Tabs are used to similar group commands. You can perform basic functions on the home page, including sorting, locating specific data inside the spreadsheet, and formatting data to make it more presentable.
2. **Ribbon bar**: The bars are used with similar group commands. For example, the Alignment ribbon bar is used to coordinate all of the commands needed to align with the data.
3. **Ribbon start button**: It is where you'll find commands like copying, saving work, creating new papers, and accessing Excel's customization options.

4.3 Microsoft Excel Environment Customization

You may create your theme. You will make the theme color appear blue if blue is your favorite color. If you are not a developer, you may not want to use ribbon tabs like a programmer. Customizations have made much of this possible. It is where you'll find commands like copying, saving work, creating new papers, and accessing Excel's customization options.

4.4 Customization of ribbon

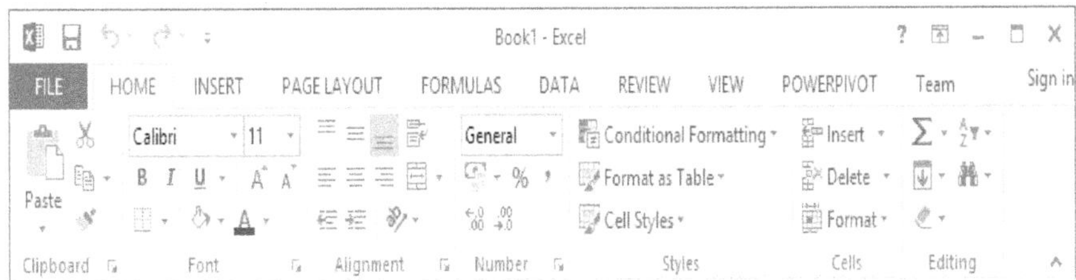

In Excel, the default ribbon looks like the one in the example above. Let's begin by customizing the ribbon. Let's assume you do not want to see some of the tabs on the ribbon, or you want to include any tabs that aren't there, including the developer tab. The options window can be used to do this.

1. On the ribbon, click the start button to bring up a drop-down menu.
2. Choose from the options on this menu. An Excel Options dialogue box will appear.
3. Pick the customize ribbon option from the left-hand side panel, as seen in the image below.

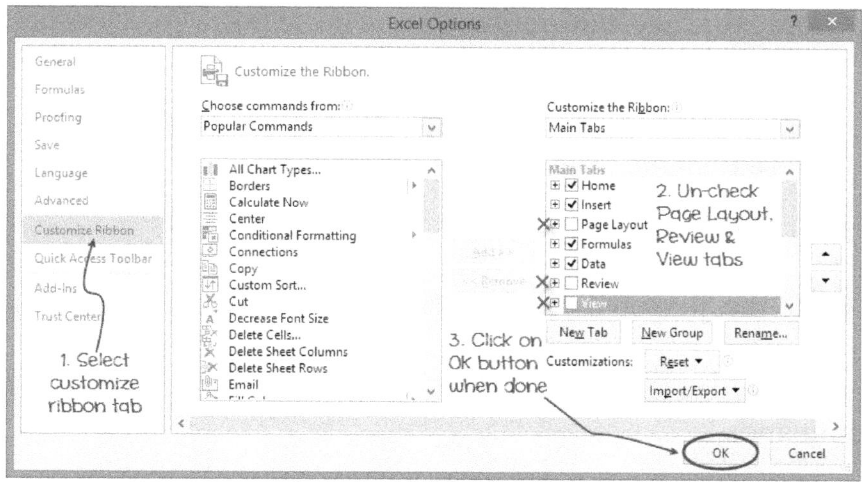

4. Remove the checkmarks from the tabs on your right-hand ribbon that you do not like. The Review, Page Layout, and Display tabs have been removed in this illustration.

5. When you are done, press the "OK" button.

The ribbon will look like this

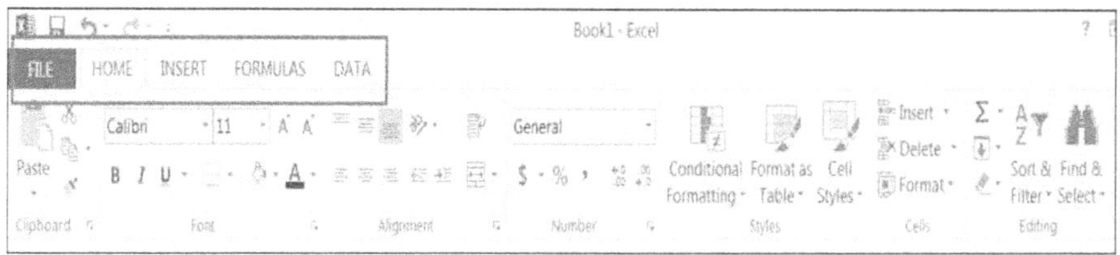

4.5 Adding custom tabs to the ribbon

You may also create a custom tab, name it something different, and assign commands to it.

1. Right-click on the ribbon and choose to customize the Ribbon. A dialogue box would pop up.

2. Choose the 'new tab' option.

3. Go to the tab you just made and choose it.

4. Choose Rename from the drop-down menu.

5. Give it a personal name.

6. Go to the 'personal' tab and choose New Group.

7. Press the Rename button and give it the name My Commands.

8. Attach commands to the ribbon bar now.

9. On the middle panel, you'll see a list of commands.

10. Choose All Chart Categories from the drop-down menu, then press the Add tab.

11. Choose OK.

The ribbon will look like this

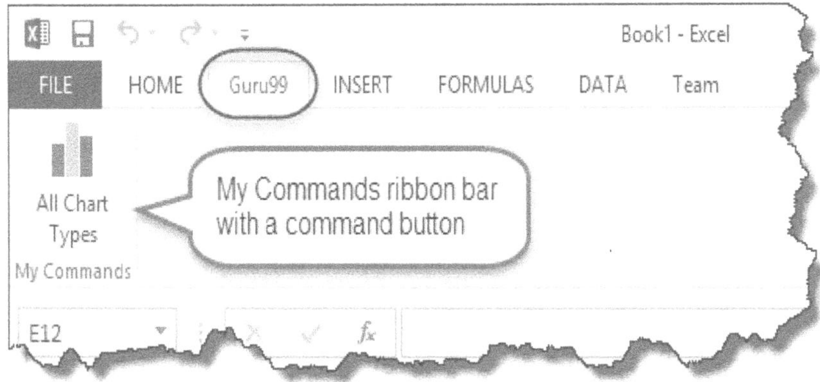

4.6 Choosing a color theme

Go to the Excel ribbon and pick the File Option command to adjust the color theme of an Excel sheet. It is going to open a window. You'll need to follow the steps below in this window.

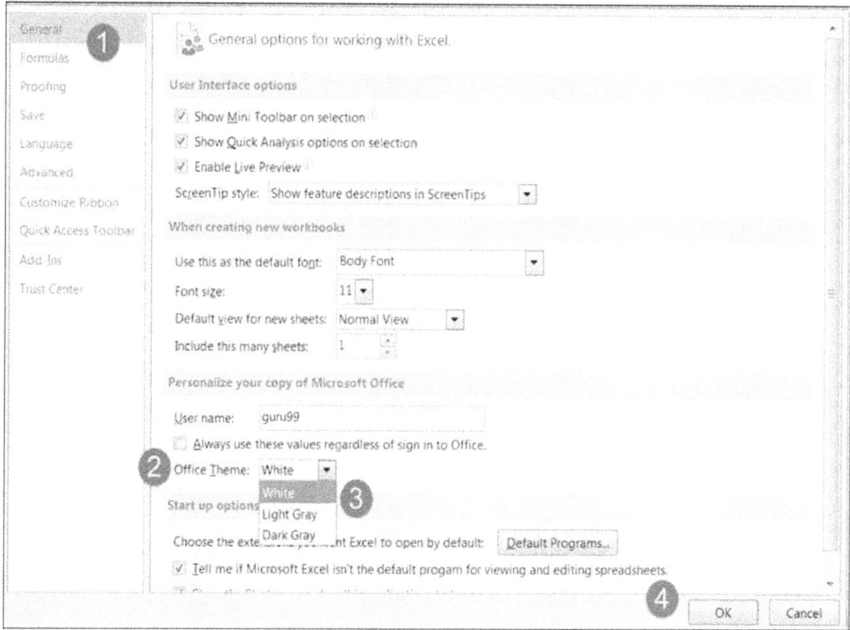

1. The general tab would be chosen by design on the left-hand column.

2. To function in Excel, go to General options to select a color scheme.

3. Pick the desired color from the drop-down list of color schemes.

4. Choose OK.

4.7 Settings for formulas

You may use this option to modify Excel's activity while working with formulas. It could be used to change the cell referencing type, use autocomplete when entering formulas, and use numbers for both rows and columns, among other items.

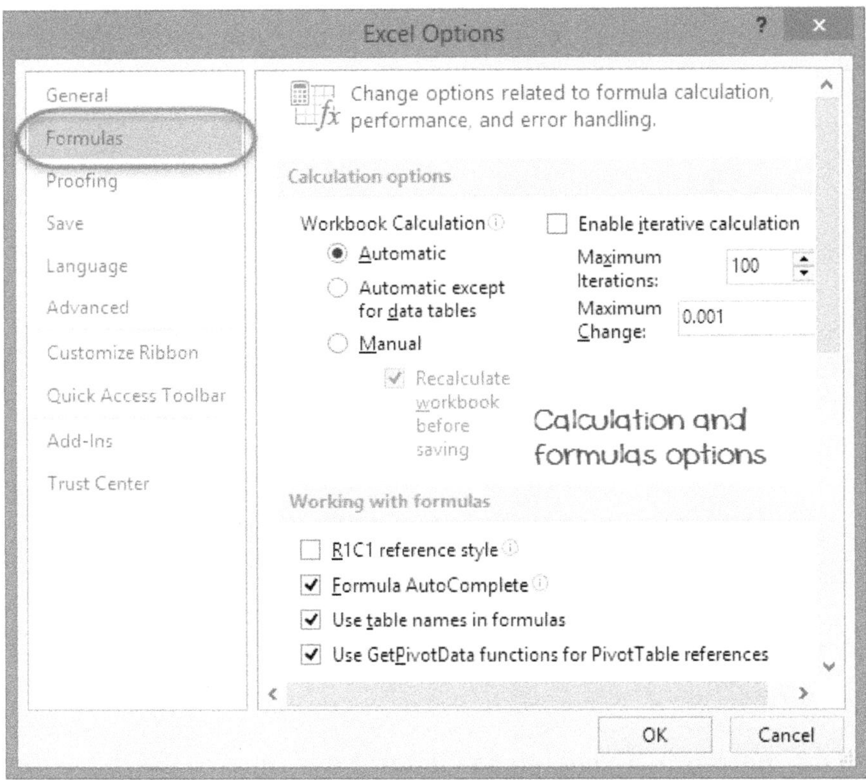

To make an option active, check the box next to it. To disable an alternative, remove the checkmark from the checkbox. This alternative can be found on the left-hand side panel of the Dialogue window, under the Formulas tab.

4.8 Proofing settings

This option modifies the text in Excel that has already been entered. It allows you to customize features such as the dictionary language used when checking for misspellings, dictionary suggestions, and so on. This option is found under the proofing tab in the options dialogue window on the left-hand side screen.

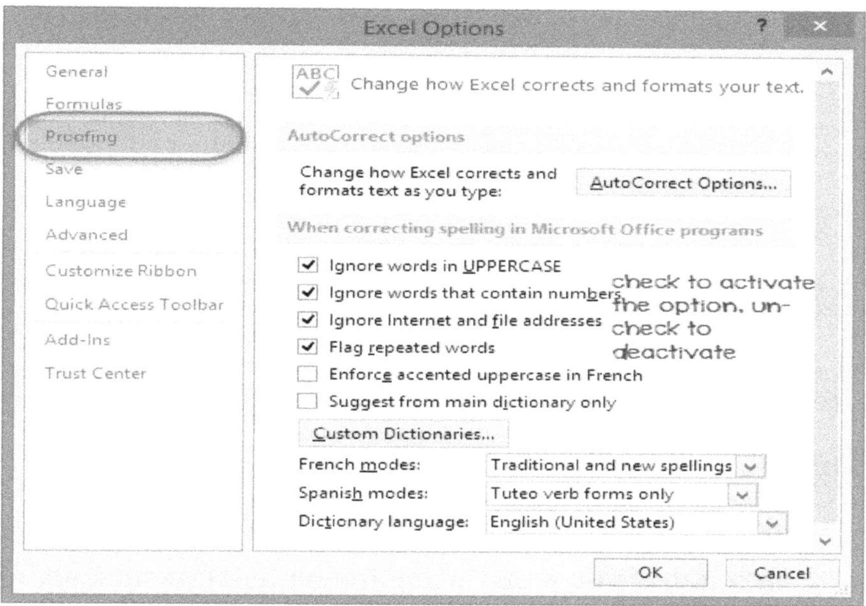

4.9 Save settings

You may use this feature to set the default file format for saving data and to allow auto-recovery if your device shuts down before you can save your work. This option is found under the Save tab on the left-hand side panel of the Options dialogue window.

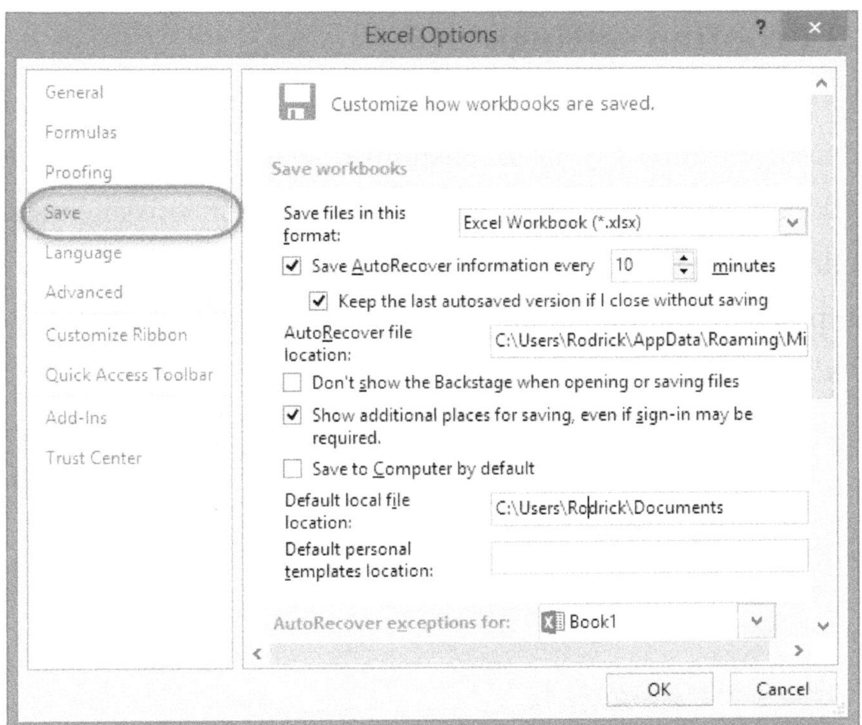

4.10 Best Practices when working with Microsoft Excel

1. When saving workbooks, keep backward compatibility in mind. If you aren't using the newest functionality of newer versions of Excel, you should save the files in 2003 *.xls format for backward compatibility.

2. Use detailed names for columns and worksheets in a workbook.

3. Formulas with a variety of variables can be avoided. Split them down into smaller, more manageable outcomes that you can draw on.

4. Where necessary, utilize built-in functions rather than writing your formulas.

4.11 Inserting Symbols

The user may access the complete character set for each font installed on their computer by clicking (Insert Tab, Symbol). This dialogue box makes it simple to insert special symbols and characters into cells. It is important to remember that each font has its collection of unique characters. All fonts do not support the Euro. Fonts that back the Euro include Tahoma, Courier, Times, and Arial.

Symbols tab

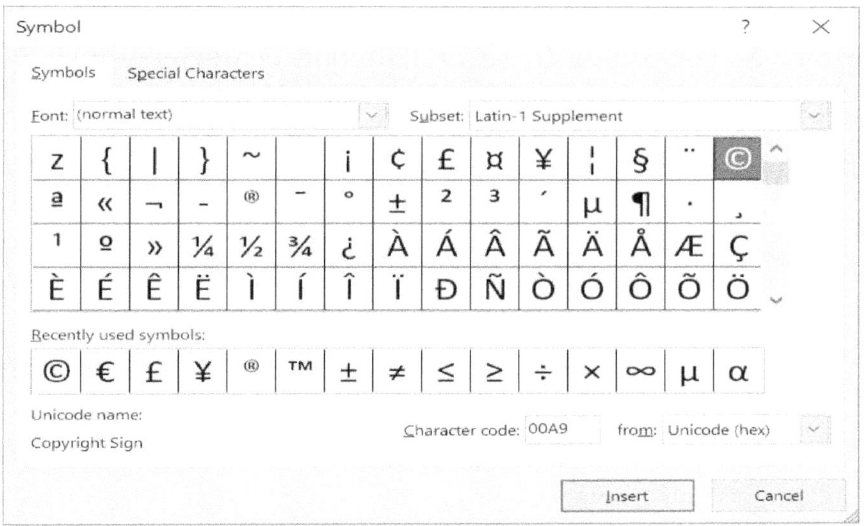

Font: Choose a font to use for displaying the symbol or character.

Subset: This function allows you to leap to specific areas inside the character set:

Greek and Coptic, Latin Extended-A, Latin-1 supplement, Basic Latin, Latin Extended-A, Latin Extended-B, Latin Extended-B, Basic Latin, Latin Extended-A, Latin Extended-B, Basic Latin, Latin Extended- General Punctuation, Currency Symbols, Letter like Symbols, Mathematical Operators, Geometric Shapes, Private Use Region, Spacing Modifier Letters

Character Code: The currently selected character's code is shown. The code can be shown in three ways: Unicode hex, ASCII decimal, or ASCII hex.

Special Characters tab:

This gives you rights to several unique characters, such as trademarks and copyright symbols. These characters will be left-aligned by default.

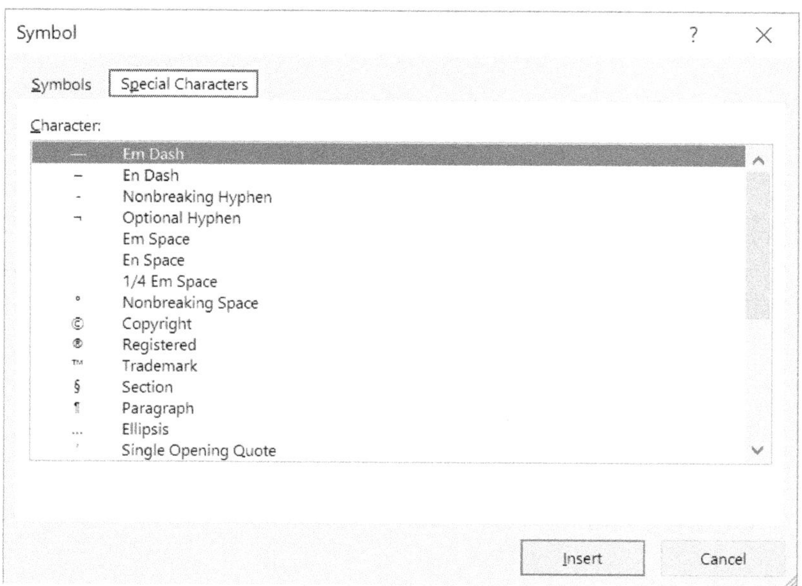

Alt + Number Pad

Alt +	Symbol / Description
21	Section - §
20	Paragraph - ¶
0128	Euro - €
0174	Registered - ®
0169	Copyright - ©

0153	Trademark - ™
0178	Subscript 2 - ^{2}
0185	Subscript 1 - ^{1}
0179	Subscript 3 - ^{3}
0189	Half - ½
0188	Quarter - ¼
0190	Three Quarters - ¾

Greek Letters

Alt +	**Symbol / Description**
224	Alpha – à
226	Gamma - â
225	Beta – á
235	Delta – ë

233	Theta – é
238	Epsilon – î
227	Pi – ã
228	Uppercase Sigma - ä
230	Mu – æ
229	Lowercase Sigma (or use Shift + F3 to change case and select upper case) - å
232	Uppercase Phi – è
231	Tau – ç
234	Omega - ê
237	Lowercase Phi – í

4.12 Understanding the worksheet (Rows, Columns, Workbooks, Sheets)

A worksheet is a set of rows and columns. When a row and a column overlap, a cell is formed. Cells are where data is held.

Each cell has its address, which is used to mark it. Columns are labeled with letters, and rows are labeled with numbers.

The term "workbook" refers to a compilation of worksheets. By default, an Excel workbook has three cells. You can add or remove sheets as required to suit your needs. The default names for the sheets are Sheet1, Sheet2, and so on. It would help if you changed the names of the sheets to something more meaningful, such as Daily Expenses or Monthly Budget.

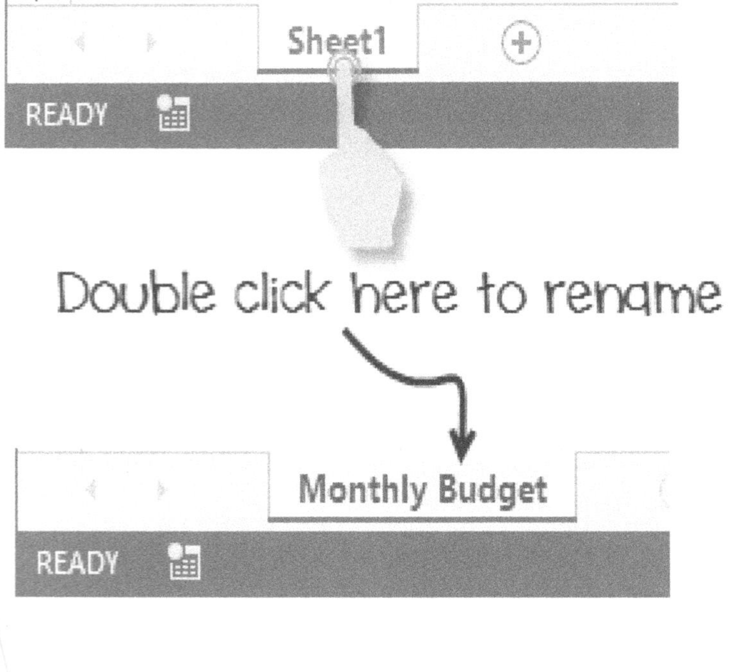

Chapter No: 5 Uses of Microsoft excel

To the common citizen, Excel is a number-crunching software that is often used to track household spending and measure complicated formulas for school assignments. However, the software is capable of far more and can be an extremely successful business method. The below are a few examples that firms make use of Excel in the workplace. Excel is commonly used today for all because it is very useful and saves a lot of time. It has been in operation for a long period and is updated annually with new functionality. The most remarkable feature of Microsoft Excel is that it can be utilized with any work. It is used for accounting, data management, research, inventory management, finance, corporate functions, and complex calculations, among other things. It can be used to perform statistical equations and store valuable data in maps or spreadsheets. The word alone inspires the vision of lengthy spreadsheets, complex macros, and the odd pivot table or bar graph. It is true—with more than 1 billion Microsoft Office users worldwide, Excel has established itself as the technical norm in offices worldwide for almost everything requiring the handling of large amounts of data.

However, if you believe Excel is just useful for getting you cross-eyed when you stare at many figures and financial records, you are mistaken. As Tomasz Tunguz points out, Excel has many uses in business (and beyond) that extend beyond basic spreadsheets. Indeed, the applications are seemingly limitless. There is no way for us to assemble a collection that encompasses all of Excel's potential uses (even if you were up for reading a War and Peace-sized listicle). To show the versatility and strength of everyone's favorite spreadsheet method, this book has compiled a list of different ways you should use Excel—professionally, privately, or just for fun. Normally, the main function of Excel is all about numbers. Excel simplifies sorting, retrieving, and analyzing a huge (or even a small!) amount of data.

There are several large categories to include when applying Excel for something numerically relevant. MS Excel secures the data, ensuring that no one else can view or manipulate them. MS Excel enables you to password-protect your data. MS Excel can be reached from any place. If you do not have computers, you can also operate on MS Excel via smartphone. MS Excel offers so many advantages that it has become an ingrained feature of the lives of millions of citizens. MS Excel has a plethora of tools and features that simplify work and save time. To use MS Excel to its full potential, it is essential to understand its benefits and advantages. The below are the most useful features of Microsoft Excel:

5.1 Calculation

Are you constantly doing the same calculations? Create a fully personalized calculator in your Excel by programming the most frequently encountered formulas. Thus, all you have to do is enter your digits, and Excel will calculate the response for you—no elbow grease needed. Since Microsoft Excel is mainly a spreadsheet application, it is incredibly efficient and robust when computing numbers and solving engineering and math problems. It allows you to total or average a column of numbers quickly. Apart from that, you can calculate weighted averages and compound interest, optimize the promotional expenditure, reduce shipping costs, and create the best job schedule for your workers. This is accomplished by entering calculations into cells. There is absolutely nothing that Microsoft Excel cannot do when it comes to equations, from totaling a column of numbers to solving some complicated linear programming problems. Microsoft excel can do everything. Excel contains a few hundred predefined formulas referred to as Excel functions to do this. Additionally, you may use Excel as a calculator to perform mathematical operations such as adding, dividing, multiplying, and subtracting numbers and raising to the power and finding roots.

5.2 Accounting

Microsoft Office Excel was created to assist with accounting tasks such as budgeting, financial statement preparation, and balance sheet formation. It includes simple spreadsheet functionality as well as several functions for conducting complicated mathematical calculations. Additionally, it allows various add-ons for modeling and financial forecasts. It integrates easily with external data, allowing you to import and distribute financial and banking data to and from other accounting software systems. Budgeting tools, forecasting tools, financial reporting, expense tracking, and loan calculators are all accessible. Excel was created to address these various accounting requirements. And, given that 89 percent of businesses depend on Excel for various accounting functions, it fits the bill. Excel also has a variety of various spreadsheet models to simplify many of those processes.

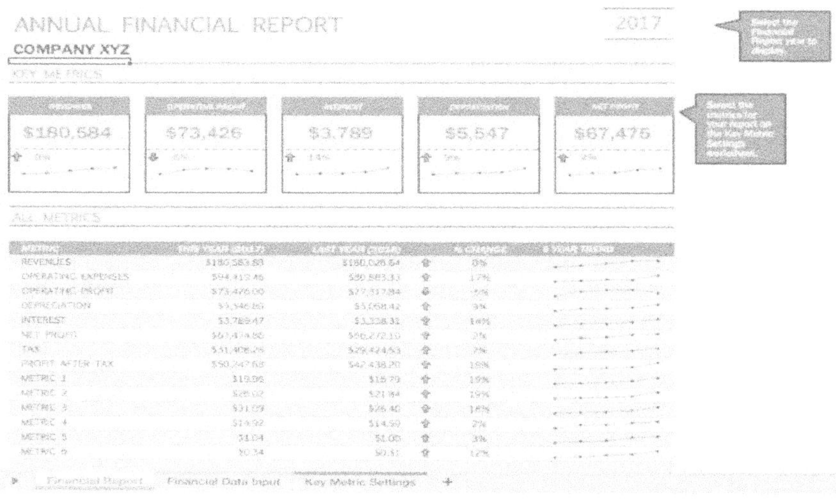

5.3 Data entry and storage

Excel is an outstanding platform for data entry and retrieval at the most simple stage. Indeed, the size of an Excel file is restricted only by your computer's processing capacity and memory. At most, 16,384 columns and 1,048,576 rows can be used in a worksheet. Thus, Excel is capable of storing a large amount of data. Not just that, functionality such as Data Form simplifies the process of entering and viewing data, allowing users to build personalized data entry forms suited to their specific business needs. Which may be used to create and manage mailing lists for customers or employee work shifts.

5.4 Charting

Charts present data in a graphical style, enabling you and the viewers to interpret the relationships between the data. When creating a chart, you can choose from various chart formats (for example, a 3-D exploded pie chart or a stacked column chart). The collection of chart types includes pie charts, line charts, scatter charts, bar charts, region charts, and column charts. Suppose you are looking for a digestible and more visual way to represent results. In that case, Excel's ability to translate rows and columns of digits into elegant charts is certain to become one of your favorite features.

5.5 Inventory tracking

Inventory tracking may be a hassle. The most straightforward method of using Excel as a stock management framework is to group your data by sales quantity. This enables you to build a flexible inventory tracking system that notifies you when it is time to order products. Fortunately, Excel will assist employees, company owners, and even people in staying prepared and on top of their inventory—even before big issues arise. However, effectively using Excel for online inventory control takes time, and properly configuring the initial template is critical.

5.6 Calendars and schedules

If you are comfortable with code, Microsoft provides a programming language to assist you in creating an excel calendar. Do you need assistance in creating a material calendar for your website or blog? Are you looking for lesson plans for your classroom? A calendar of paid time off for you and your coworkers? Do you want a regular schedule for you or your family? Excel is remarkably stable when it comes to managing several schedules. Utilizing a calendar template is convenient. All you have to do is choose if you require a monthly or annual calendar and fill the template with your scheduled activities. Additionally, you can customize the fonts, font sizes, and colors. If you use the monthly calendar, you must first modify the title and dates for the month you choose to use.

Daily Schedule

Week: [Date] Start Time: 5:00 AM

	Mon	Tue	Wed	Thu	Fri	Sat	Sun
5:00 AM		Go to gym					
5:30 AM							
6:00 AM							
6:30 AM							
7:00 AM							
7:30 AM							
8:00 AM							
8:30 AM							
9:00 AM							

5.7 Business data collection and verification

Businesses also use several processes (CRM, inventory management, etc.), each with its logs and database. All of which can be easily exported to Excel. Additionally, the software should be used to clean up data by deleting redundant or incomplete entries; removing those data from the start is important because it can impact subsequent analysis and reporting. Utilizing Excel for commercial purposes will assist you with data processing activities. You should structure the content of worksheets so that data appear accurately on the computer and when printed. You should design data to assist readers in locating and comprehending particular categories of information.

Personal Inventory		
Name [Name]	Insurance Company [Company]	Agent Address [Address]
Address [Address]	Agent [Name]	Agent Phone [Phone/Fax]
Phone [Phone]	Company Phone [Phone]	Agent Email [Email]
Email [Email]	Policy Number [Policy]	

Item Description	Category	Serial Number	Value
[Item]	[Category]	[Serial #]	[Value]
[Item]	[Category]	[Serial #]	[Value]
[Item]	[Category]	[Serial #]	[Value]
[Item]	[Category]	[Serial #]	[Value]
[Item]	[Category]	[Serial #]	[Value]
[Item]	[Category]	[Serial #]	[Value]
[Item]	[Category]	[Serial #]	[Value]
[Item]	[Category]	[Serial #]	[Value]
Total			**$0.00**

5.8 Making a plan

Let's move on from the numbers, Excel will assist you in planning and organizing various tasks that do not include countless rows of digits. Excel enables experts to analyze and represent large amounts of data using graphs and charts. Easily track and report the status of the projects. You may create timeline charts and graphs to assist with tracking the progress of your project. Templates for project management that assist in the collection of data and the identification of dependencies.

5.9 Seating charts

Since seating charts assist in employee coordination, you can find it beneficial to build a company directory in the first sheet of the Excel file to compile everyone's information and contact information in one location. Moreover, if you plan a big business luncheon or a wedding, creating a seating chart may be a real hassle. Fortunately, Excel will simplify the process considerably. If you are a total whiz, you'll be able to build your seating chart automatically using your RSVP spreadsheet.

5.10 Goal planning worksheet

From professional and health and financial goals, having something to keep you centered and on track is beneficial. Attaining large targets is not easy, so the majority of people fail to achieve the majority of theirs. However, with the proper context, almost everything is possible. Unfortunately, creating a to-do list and hoping for the best is not an option. Here is how to go from target setting to goal accomplishment. What you need is your preferred task management application, a spreadsheet, and a schedule. Excel is a work of art. The platform enables you to build a variety of worksheets, planning documents and logs that will assist you in tracking the progress—and, ideally, crossing the finish line.

Task	Times/Week	S	M	T	W	T	F	S	Complete
Go for a run	2	✔		✔					Yay!
Don't Leave Dirty Dishes Overnight	2	✔			✔		✔		Yay!
Eat 1 Fruit or Vegetable	3	✔			✔	✔	✔	✔	Yay!
Floss	3	✔		✔			✔	✔	Yay!

8-3-2014

Task	Times/Week	S	M	T	W	T	F	S	Complete
Go for a run	2								0 of 2
Don't Leave Dirty Dishes Overnight	3								0 of 3
Eat 1 Fruit or Vegetable	3								0 of 3
Floss	3								0 of 3

8-10-2014

Task	Times/Week	S	M	T	W	T	F	S	Complete
Go for a run	3								0 of 3
Don't Leave Dirty Dishes Overnight	3								0 of 3
Eat 1 Fruit or Vegetable	4								0 of 4
Floss	4								0 of 4

8-17-2014

Task	Times/Week	S	M	T	W	T	F	S	Complete
Go for a run	3								0 of 3
Don't Leave Dirty Dishes Overnight	3								0 of 3
Eat 1 Fruit or Vegetable	5								0 of 5
Floss	4								0 of 4

5.11 Managerial and Administrative duties

Apart from recordkeeping, Microsoft Excel is useful in office management for helping day-to-day activities such as invoicing, bill payment, and communication with vendors and customers. It is a general-purpose application for tracking and coordinating office tasks. Creating and detailing company procedures is one of the managerial responsibilities. This assists in process optimization and is an efficient method of arranging processes and scenarios. Excel has resources for creating flow charts that may contain text, images, and animations.

5.12 Mock-ups

A mockup, or mockup, is a full-size model or a scale of a concept or product used in manufacture and design for teaching, presentation, design assessment, and promotion, among other uses. A mockup is considered a prototype if it demonstrates at least some of the features of a system and allows testing of the design. Excel might not be the first application that comes to mind when considering design. However, believe it or not, the platform may be used to create a variety of mockups and designs. Indeed, it is an unexpectedly common option for wire framing and dashboard design.

5.13 Getting stuff done

Microsoft Excel is one of the most popular programming options – due to the critical role in a variety of sectors. It is the most widely used spreadsheet software in various business and academic practices and personal data organizations. So do you want to ramp up your productivity? Excel will intervene to save the day with several functions that can assist you in completing the projects and to-dos with simplicity and organization.

5.14 Task list

One of the most compelling reasons to maintain a to-do list is for organizational purposes. Using a list to organize your activities will help you feel more in control and grounded. Having a concise overview of accomplished and unfinished activities will assist you with being orderly and mentally concentrated. Say goodbye to the traditional paper and pen to-do list. With Excel, you can create a much more comprehensive task list and track the progress on the larger tasks currently on your plate.

TASK LIST

MY TASKS	START DATE	DUE DATE	% COMPLETE	DONE	NOTES
[Task]	[Date]	[Date]	0%		
[Task]	[Date]	[Date]	50%		
[Task]	[Date]	[Date]	100%	●	

5.15 Checklist

If you are creating a spreadsheet to share with others or simply for personal tracking, using a checklist in Microsoft Excel can make data entry a breeze. You can create a simple checklist that allows a person to check off items they've purchased or completed—from a grocery list to a list of to-dos for an upcoming marketing campaign.

PURCHASED?	GROCERIES:	
☐	Apples	
☐	Tomatoes	
☐	Milk	
☐	Eggs	
☐	Cheese	
☐	Bread	

5.16 Accounting and budgeting

Excel also provides easy-to-use budgeting and accounting templates. From there, you can use the software's built-in formula and calculating features to assist you in organizing and synthesizing data. Microsoft Office Excel includes making forecasts, cash balance statements, and profit-and-loss statements, three of the most fundamental accounting documents. Additionally, you can import more complex statement and budgeting templates from the Office platform, or you can buy and install customized templates from third-party suppliers. Suppose you want to create complex or customized financial statements or budgets. In that case, you can modify a current design and reuse its components or start from scratch utilizing Excel's built-in functionality.

5.17 Project management charts

This topic has already mentioned how Excel is a tremendous beast when it comes to chart development. Additionally, this concept is true when it comes to different project management charts. Excel can be used to construct a digital timeline chart and to assist you in mapping out a project's timetable and phases. Specifically, you should build a Gantt chart, a common project management method, since it organizes projects according to their length, start date, and completion date. There are several aspects that Excel will assist you with keeping your project on schedule, from waterfall charts to kanban type boards (very much like Trello!) to keep stuff organized.

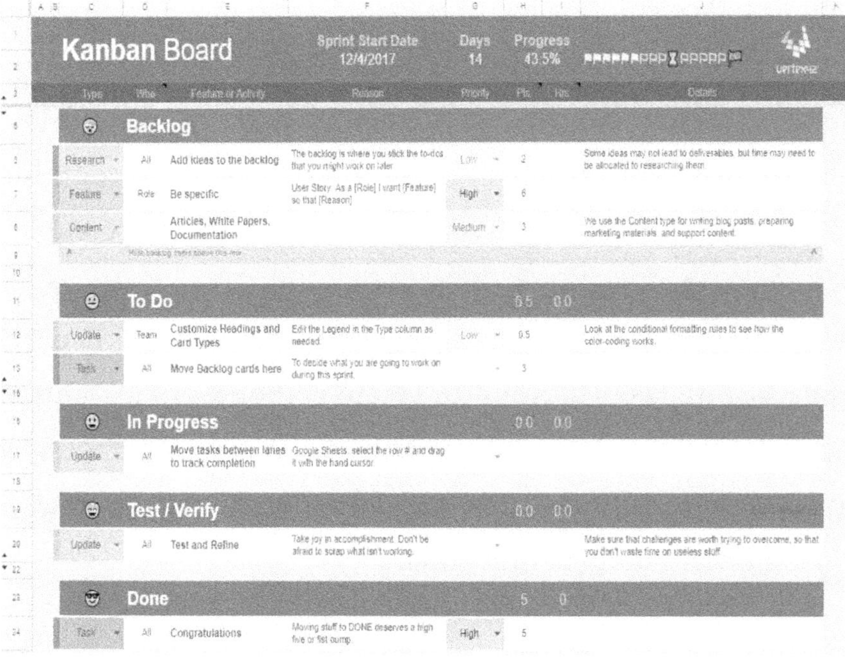

5.18 Time logs

Keeping an inventory of employee time in an Excel database is not the simplest way to manage payroll hours. Consider automatic time monitoring that works easily with the payroll solution for this purpose. However, if you are still on the fence about whether spreadsheets are good for you, you can use one of the free timesheet models or learn how to build an Excel timesheet. You are aware that time management can be a tremendous benefit to you and your efficiency. Although there are several sophisticated applications and software to assist with this task, you should think of Excel as the initial time-tracking method. And it remains a viable alternative today.

Time Sheet

[Employee name] | [Email] | [Phone]

Manager | [Manager name]

Period [Start date] - [End date]

Standard Work Week	Hours Worked	Regular Hours	Overtime Hours
40.00	0.00	0.00	0.00

Date(s)	Time In	Lunch Start	Lunch End	Time Out	Hours Worked
[Date]	[Time In]	[Lunch Start]	[Lunch End]	[Time Out]	0.00
[Date]	[Time In]	[Lunch Start]	[Lunch End]	[Time Out]	0.00
[Date]	[Time In]	[Lunch Start]	[Lunch End]	[Time Out]	0.00
[Date]	[Time In]	[Lunch Start]	[Lunch End]	[Time Out]	0.00
[Date]	[Time In]	[Lunch Start]	[Lunch End]	[Time Out]	0.00

5.19 Creating Forms

Excel has a plethora of valuable features for data entry. Among them is the Data Entry Form. Data entry may be a significant aspect of utilizing Excel at times. It may be difficult for the data inputted to know where to place which data with nearly infinite cells. A data entry form will address this issue and assist the user in entering the necessary data in the correct location. Excel is an excellent tool for generating forms of all types, from basic to complex. Additionally, you should program different drop-down menus such that users can browse from a pre-defined catalog.

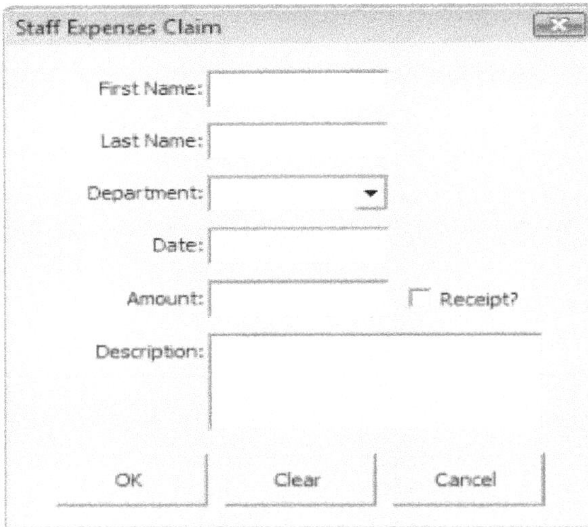

5.20 Data Analysis

So, you've been handed a mountain of data and tasked with deriving information from it. Not to panic, Excel will also assist you in managing and synthesizing simple and infectious data.

Pivot Tables are one of the best features for accomplishing this. They enable users to consolidate and concentrate on specific data segments within a vast data collection, resulting in concise snapshots that can be used as an immersive summary report. The table can be easily customized to represent preferred data fields by adding filters or switching out data segments.

5.21 Quizzes

Are you attempting to test another person's — or even your own — understanding of a subject? You can build a bank of answers to questions in Excel and then advise Excel to test you in another worksheet. Creating question and answer quizzes is an excellent opportunity to practice your Excel skills while still having fun. Excel is an excellent platform for creating quizzes for work or play. It will keep track of right and incorrect responses and maintain a running score of your success. You can create your own set of questions or take them from your course book and vice versa

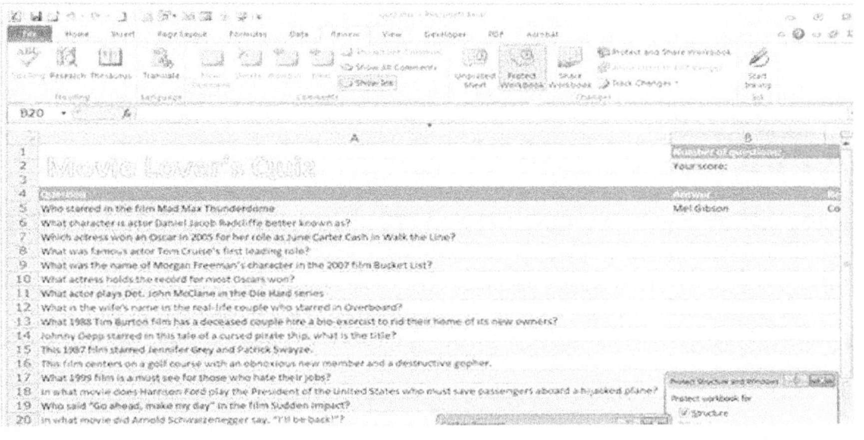

CRM

Are you looking for a lightweight CRM that can help you keep top of mind with your customers? One can be generated in Excel. And, what's more, the best part? Through creating your own, you ensure that everything is fully customizable. Additionally, Sales Hacker has compiled a handy set of free sales excel models to assist you in getting started! The CRM Excel template is a simple spreadsheet for arranging leads and contacts. It contains a correspondence log that keeps track of the last message, follow-up activities, the next contact's date, and lead status.

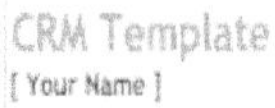

Name	Company	Work Functio	Phone	Email	Estimated Sale	Last Contac	Next Actio	Next Contac	Lead Statu	Lead Sourc	Notes
Jameson, Bill	XYZ Plumbing	Owner	444-555-6666	xyz@plumber.com	$ 45,000	1/10/13		1/29/13	Cold	Referral	
Anderson, Jane	ABC Corp	Sales Manager	222-656-7890	bus@abccorp.com	$ 10,000	1/25/13		2/5/13	Warm	Website	
Smithers, Joe	ACME	Business Dev.	111-234-5678	acme@acme.com	$ 4,500	1/27/13		2/15/13	Active	Email	Loves chocolate

Insert new rows above the gray line

5.22 Reporting and Visualizations

Charts and graphs can be created using data from both pivot tables and raw data sets, which may be used to create structured reports, presentations, or assist in analyzing data, Because they may have a different viewpoint on patterns and performance.

Excel has a similar selection of pre-designed map models but enables users to customize information such as axis values, colors, and text comments. Visual reporting applies to all business sectors. For example, marketing departments may use a column chart to focus on the effectiveness of an advertising strategy over time and in comparison to prior campaigns. Excel allows you to view the data analysis reports in a variety of ways. However, suppose the data collection findings can be visualized in maps that show significant points in the data. In that case, the viewer can easily understand what you are attempting to convey with the data. Additionally, it has a positive effect on the presenting style.

5.23 Mailing list

Data does not have to be numerical. Excel is also excellent at handling and sorting huge lists of names and addresses, making it the ideal tool for the company's holiday party invite list or mailing list for a large promotion or campaign.

	A	B	C	D	E	F
1	FIRST NAME	LAST NAME	ADDRESS	CITY	STATE	ZIP CODE
2	Oprah	Winfrey	123 Magnificent Mile Ave.	Chicago	IL	58922
3	Mister	Rodgers	8935 Beautiful Day Rd.	New York	NY	23935
4	Hulk	Hogan	9284 Hollywood Blvd.	Los Angeles	CA	39825

Additionally, you can mail merge using Excel, which simplifies the process of printing address labels and other required materials. Additionally, you may use a similar concept to build folders, RSVP lists, and other rosters that provide a large amount of information about individuals.

5.24 Historical logs

If you want to keep track of the various craft breweries you've tried, the exercises you've done, or anything else, consider Excel to be your go-to resource for organizing and logging such items. Numerous accountants and analysts are used to analyze patterns based on historical and present results to make accurate predictions for the future. It is a natural occurrence that often results in consternation and bad Excel modeling. It may, however, be circumvented by the use of a basic interface and the implementation of functions in Microsoft excel.

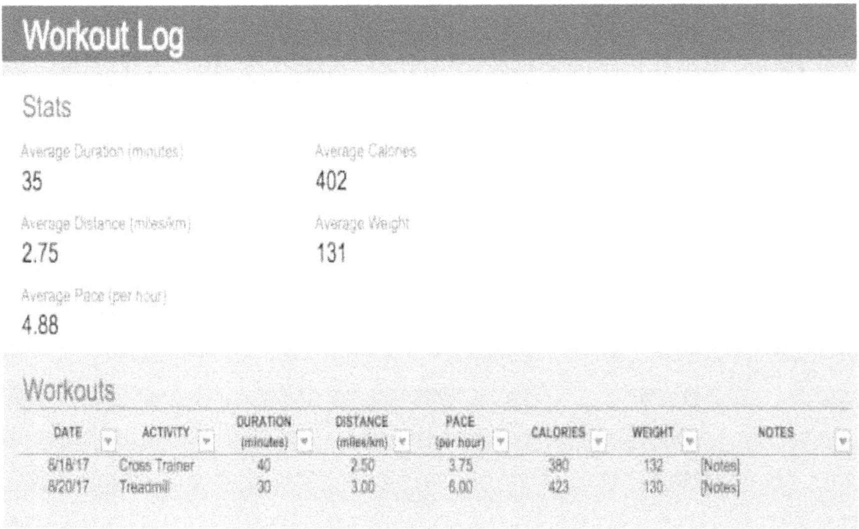

Workout Log

Stats

Average Duration (minutes)
35

Average Calories
402

Average Distance (miles/km)
2.75

Average Weight
131

Average Pace (per hour)
4.88

Workouts

DATE	ACTIVITY	DURATION (minutes)	DISTANCE (miles/km)	PACE (per hour)	CALORIES	WEIGHT	NOTES
8/18/17	Cross Trainer	40	2.50	3.75	380	132	[Notes]
8/20/17	Treadmill	30	3.00	6.00	423	130	[Notes]

5.25 Art and animations

When you consider Excel, you are inevitably reminded of statistics, estimates, and finances. It is highly improbable that you'll ever associate spreadsheets with fun, imagination, or animation, but maybe we can change that! People are using spreadsheet tools as a painting medium to create incredible works of visual art. Excel does not have to be extremely practical. There are many other enjoyable items that you can build with the spreadsheet app.

Compared to costly imaging tools such as Photoshop, Excel (or Google Sheets or various other spreadsheet applications) is often pre-installed on your computers. It conceals an unexpected amount of creative functionality within its toolbar, rows, and columns. Excel's features are almost certain to exceed the original expectations. Indeed, several users have used the tool to produce some really amazing artwork, ranging from portrait images to animations.

5.26 Sudoku puzzles

Are you a fan of Sudoku puzzles? If it happens, you will build your own using Excel. Or, if you are stuck on an especially difficult one, you should enlist the assistance of Excel to help you!

	A	B	C	D	E	F	G	H	I
1					1	7			
2	4						5		
3			9						
4						1		9	
5			5	7	8				
6	1			2					6
7	5		4		2				3
8	2				3	8		6	4
9		1	3			9		5	

5.27 Word cloud

A word cloud is a graphical depiction of the words used in a passage of text or sequence of passages. Tags are usually single words, and the font color and scale of each word indicate its value. It is an excellent method of revealing critical information. Another benefit of the word cloud is that it is easy to build using our Excel Add-In. Thus, you save time by not having to analyze each statement independently. Automated processes will take care of things for you in a matter of minutes. Not only are word clouds more physically appealing than a table of details, but they are also very simple to comprehend. Creating word clouds in Excel is a basic but effective technique for visualizing the relevance of repeated terms and phrases in a variety of unstructured text sources, including social networking, product feedback, consumer and employee surveys, and more. Perhaps word clouds are not the most scientifically accurate representations of results. However, they are an enjoyable (not to mention wonderful) way to appreciate the most often used words. You guessed it—using Excel, you can make one.

5.28 Manage expenses

MS Excel assists with expense management. If a doctor earns about 50,000 per month, he can experience certain costs, and if he wishes to determine precisely how much he spends per month, he will comfortably do so using MS Excel. He will enter his monthly revenue and expenditures in excel tables and determine how much he is investing and manage his expenses appropriately.

There are some advantages of using MS Excel; this is why it is used by people worldwide for various activities. It not only saves time but also facilitates work. It is capable of almost any mission. For instance, you can perform mathematical calculations and create graphs and charts to store the data. Calculations and data storage is simple for the businessman. MS Excel allows you to store and review vast amounts of data.

It assists in consolidating data in one location, ensuring that data is not lost and that time is not wasted searching for specific data. As a result of these reasons, it has grown in popularity, and people have developed a habit of using it.

5.29 Forecasting

When you make a forecast, Excel generates a new worksheet with a table containing the historical and expected values and a chart expressing the data. A forecast will assist you in forecasting potential revenue, inventory needs, and consumer trends. While monitoring and updating performance is critical for any company, planning and being prepared for various scenarios and adjustments is equally crucial. Excel may be used in combination with third-party software to simulate financial forecasts using historical data. Additionally, Excel will use the data from a table to construct a calculation used to measure future values.

5.30 Trip planner

Are you unsure how to schedule a trip? Are you planning a vacation? Suppose you are preparing a cruise, a weekend getaway, a long-term sabbatical, a road trip, a ski break, or an adventure as a couple, band, party, or alone. In that case, our guide walks you through the stages of planning, from motivation to saving to preparing to survive on the road. Be certain you've addressed everything by making a supportive itinerary before packing your bags and departing. Excel also has a convenient travel planner template that will ensure you do not forget everything (from the budget to airline information!).

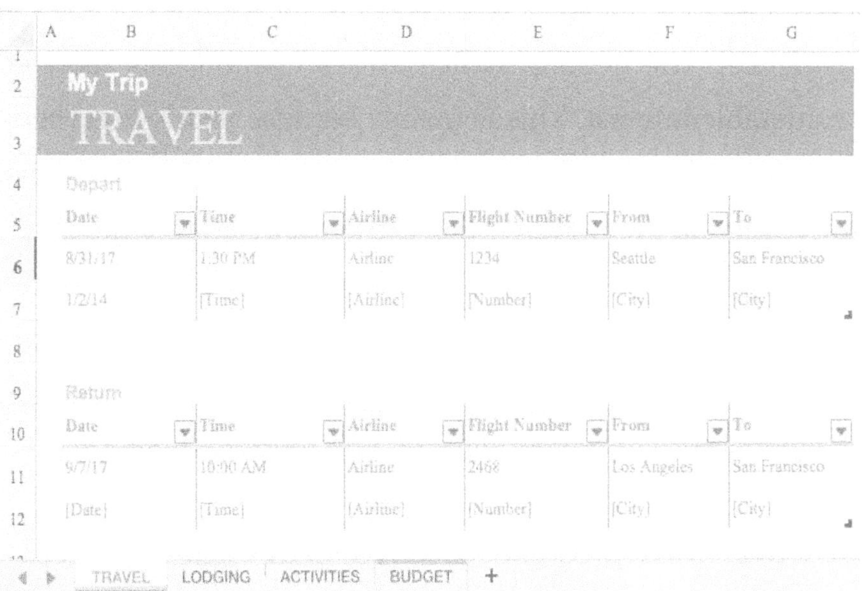

Chapter No: 6 Microsoft office 2021

Excel is a spreadsheet application that comes as part of the Microsoft Office suite. It is often used to build grids of formulas and numbers that include information about calculations, product monitoring, and accounting, among other things. As you are no doubt aware, Microsoft enhanced Excel 2019 with a slew of new features. Additionally, the tech giant continues to upgrade the venerable spreadsheet program through Office 365 updates. Of course, these latest features are useless until you are mindful of them and understand how to utilize them. Microsoft is introducing a few notable updates to Office 2021. Can the latest upgrades make a difference in terms of the overall experience? The news of Office 2021's opening has received considerable interest. This is mainly because nearly everybody now uses at least one of the many Microsoft Office systems. Thus, what did they modify? Can the latest updates to the Office package improve the overall experience with the suite's services, or will they make matters significantly more complicated? If you are curious about the changes that Microsoft Office 2021 would bring, just continue reading.

6.1 Converting to a one-time-purchase model

One of the most significant uproars was undoubtedly the shift in the buying model. With the 2021 Office, you'll get two years of complimentary service before being charged on a monthly basis. Microsoft has announced that Office 2021 will be supported for five years under the one-time purchasing model. This ensures that you can register for the service once and have it for five years without restriction. When you consider that before, you may use Office for about seven years before being required to pay every month, this is a significant downgrade.

Not to note that commercial users would see a 10% rise in the price of Office Professional Plus, Office Standard, and other individual applications. The price of Office 2021 for consumers and small companies remains unchanged. Microsoft Office 2021 is compatible with both macOS and Windows, and more details regarding the features will be released in the future. If you are interested in their collaboration offerings called Office 365, a subscription-based model would be available that includes items such as PowerPoint, Word, Excel, and others. This contract is $69.99 a year for individuals or $99.99 a year for families.

6.2 Two new office versions to look forward to

Microsoft has released two latest Office versions that you are sure to be interested in Office 2021 for consumers and Office LTSC for enterprise clients (Long term servicing channel). The latest edition of Office is designed for people who are not involved in subscribing to cloud-based Microsoft 365 versions. Office 2021 has some significant feature changes, but it is still close to Office 2019, so you should anticipate some familiarity.

As previously said, Microsoft has not yet announced all Office 2021 upgrades, but the Office Long-term servicing channel, or Office LTSC, version would provide compatibility enhancements, dark mode support, and additional functionality such as Dynamic Arrays and XLOOKUP in Excel. As Microsoft previously said, Office 2021 will be available to anyone who does not want to subscribe to Microsoft 365. Microsoft has agreed to stick to another perpetual edition of Office, although the pricing and service model for these future models will change.

Support for Office LTSC timing is also more consistent with how Windows is supported. Additionally, Microsoft is tightly aligning the release cycles for Windows and Office. Both Office LTSC and Windows LTSC models are scheduled to be launched in the 2nd half of this year. Their goal is to essentially link them so that organizations can deploy and manage on a common cadence. If you are curious about how all this will look, you'll be pleased to learn that Microsoft plans to resell an Office LTSC demo in April. There will, though, be no glimpse of the consumer Office in 2021. Both recent Office models come pre-installed with OneNote and are available in 64-bit and 32-bit versions.

6.3 Dark Mode simplifies things

Dark mode users suggest that it can improve the contrast between the text you read and the context. This will theoretically make reading on your computer easier. Microsoft Office has themes of black and medium grey. You may customize the appearance of Office applications such as Excel, Microsoft Word, Outlook, and PowerPoint by selecting a dark theme. The dark mode would be significantly better, making it much darker and easy on the eyes. The dark background can greatly facilitate late-night writing and editing. Currently, Microsoft Word's dark mode darkens only the document's edges, leaving the rest of the document light white. In a future update, you will be able to darken the whole document.

6.4 Other significant changes headed to Microsoft in 2021

Microsoft also revealed that it would transition some of its software to a subscription-payment model this year. The next Skype models for SharePoint Server, Business Server, and Exchange Server will also be released in the 2nd half of this year, but only as subscription license purchase. This implies that the perpetual model would no longer be accessible. Another significant shift is the introduction of Power Apps and the stand-alone licenses. Currently, user privileges are included free of charge for Office 365 and Dynamics 365.

However, Microsoft will improve usage rights next year, which may result in a price increase for Power Apps. If you have been using Power Apps since 2019, you must buy the stand-alone license. Finally, but certainly not least, the MS Open Licensing scheme has ended. For more than two decades, this scheme has allowed small and medium-sized businesses to obtain a few permanent licenses for applications at a reasonable price.

However, as a business user, you will no longer be eligible to buy or upgrade software licenses or internet facilities after January 1, 2022. Rather than that, there would be a one-time fee on a non-subscription license that will never expire. Microsoft claims the move is part of an attempt to ease and anticipate software licenses by introducing the perpetual software license purchases.

6.5 Utilize Excel's ideas feature to Automate Data Analysis

Ideas is an artificial intelligence feature built into Excel and is accessible as part of Office 365 packages. With Ideas, Excel will easily review the data to reveal insights you would not have found otherwise. For instance, ideas can be advantageous in the following situations:

1. Analyzing purchases to rate data and classify items that are substantially greater or smaller than the majority of the population

2. Doing trend analysis to identify long-term patterns in data
3. Identifying major outliers in data points, such as technically incorrect or fraudulent transactions; and
4. Highlighting cases under which a significant part of the total value is due to a particular cause.

If you are using Excel as part of an Office 365 package, you can view Ideas from the Ribbon's Home tab. Take note, however, that this function needs an active internet connection.

6.6 Using MAXIFS, IFS, and MINIFS to create simplified conditional formulas

With MAXIFS, IFS, and MINI FS, you can easily generate formulas that include several tests. Prior to the introduction of IFS, several Excel consumers often "nested" several IF functions within a single formula. This was a standard procedure where an estimate needed to be created depending on the fulfillment of one or more criteria. However, since the advent of IFS, certain formulations have become significantly simplified. There are some formulas which requires just one IFS operation to conduct three checks on the data in cell A2. This methodology is in comparison to the numerous IF functions that were previously required.

As with IFS, MINI FS and MAXIFS enable you to perform numerous tests on your data. When MAXIFS is used, Excel returns the highest value that passes both tests. In contrast, when MINI FS is used, Excel returns the smallest value that passes all tests. Excel 2019 users have access to these features. Additionally, they are accessible to Excel customers for Office 365 subscriptions.

6.7 XLOOKUP: A more powerful and user-friendly alternative to VLOOKUP

Beginning in February 2020, Microsoft will include XLOOKUP in Excel offered as part of Office 365. XLOOKUP is a better replacement for VLOOKUP and related features such as INDEX and HLOOKUP. Although these legacy features will exist in Excel, many people will prefer the simplicity and intuitiveness of XLOOKUP. The majority will also discover XLOOKUP to be more effective. Several important distinctions between XLOOKUP and other lookup functions include the following:

1. XLOOKUP uses an exact match by contrast, while HLOOKUP and VLOOKUP use an approximate match.
2. Unlike VLOOKUP and HLOOKUP, you do not need to assign a column index number or a row index number for XLOOKUP.
3. For XLOOKUP, the order of rows and columns is negligible. This is because the feature, when used in

place of VLOOKUP, can look to the right or left. Similarly, when used in place of HLOOKUP, it may look above or below.

4. XLOOKUP enables you to decide what should happen if the lookup value is not identified, without the need for an IFERROR feature.

Illustrating XLOOKUP

To begin comprehending the benefits of XLOOKUP, consider the illustration given in Figure below. In this example, XLOOKUP is used to locate the value in cell H2 within the range B3 to B15. Keep in mind that XLOOKUP uses an exact match by nature, while VLOOKUP uses an estimated match. After locating the desired value, XLOOKUP returns the value from the range D3 to D15. If there is no match, the formula returns the phrase "Item Not Identified." Due to the relative convenience of XLOOKUP in comparison to HLOOKUP, VLOOKUP, and other related features, it is possible that when more users have access to it, XLOOKUP will become the favored lookup function.

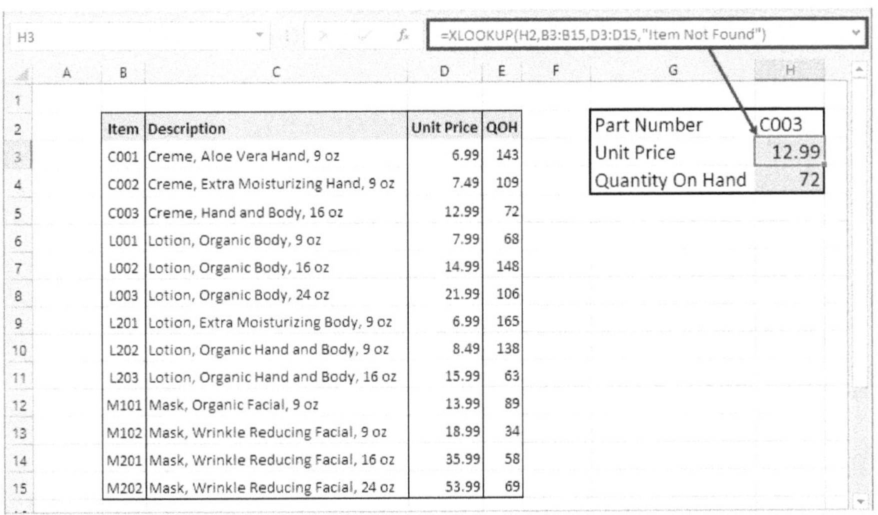

=XLOOKUP(H2,B3:B15,D3:D15,"Item Not Found")

Item	Description	Unit Price	QOH
C001	Creme, Aloe Vera Hand, 9 oz	6.99	143
C002	Creme, Extra Moisturizing Hand, 9 oz	7.49	109
C003	Creme, Hand and Body, 16 oz	12.99	72
L001	Lotion, Organic Body, 9 oz	7.99	68
L002	Lotion, Organic Body, 16 oz	14.99	148
L003	Lotion, Organic Body, 24 oz	21.99	106
L201	Lotion, Extra Moisturizing Body, 9 oz	6.99	165
L202	Lotion, Organic Hand and Body, 9 oz	8.49	138
L203	Lotion, Organic Hand and Body, 16 oz	15.99	63
M101	Mask, Organic Facial, 9 oz	13.99	89
M102	Mask, Wrinkle Reducing Facial, 9 oz	18.99	34
M201	Mask, Wrinkle Reducing Facial, 16 oz	35.99	58
M202	Mask, Wrinkle Reducing Facial, 24 oz	53.99	69

Part Number	C003
Unit Price	12.99
Quantity On Hand	72

Figure - Using XLOOKUP Instead of VLOOKUP

6.8 Dynamic Arrays

Dynamic arrays are another recent feature that is only accessible as part of an Office 365 package. With dynamic arrays, you can compose a single formula that affects several cells concurrently without duplicating the formula in each cell. Additionally, suppose you are using a variant of Excel that facilitates dynamic arrays. In that case, a user no longer needs to insert a standard array formula using the CTRL + SHIFT + ENTER keystroke series. Additionally, if you are utilizing an Excel version that facilitates dynamic arrays, six additional functions are available to help you leverage this newly acquired capability. FILTER, RANDARRY, SORTBY, SORT, SEQUENCE, and UNIQUE are among the six features.

Illustrating the Dynamic Arrays

Utilizing the latest FILTER function, given below is a basic illustration of dynamic arrays. As inferred by its name, the FILTER feature enables formula-based filtering of data in a table or range. The syntax is pretty straightforward, as shown below.

=FILTER (array (range or table), including (a Boolean array specifying which items should be included))

A third statement – [if empty] – is optional and defines the attribute to show if the filter returns null. The illustration in the Figure given below demonstrates how to use the FILTER function to filter the data without affecting the initial array. Notably, when the volume of data in the table referred to by the FILTER function expands or decreases, the volume of data returned by the calculation often expands or gets smaller.

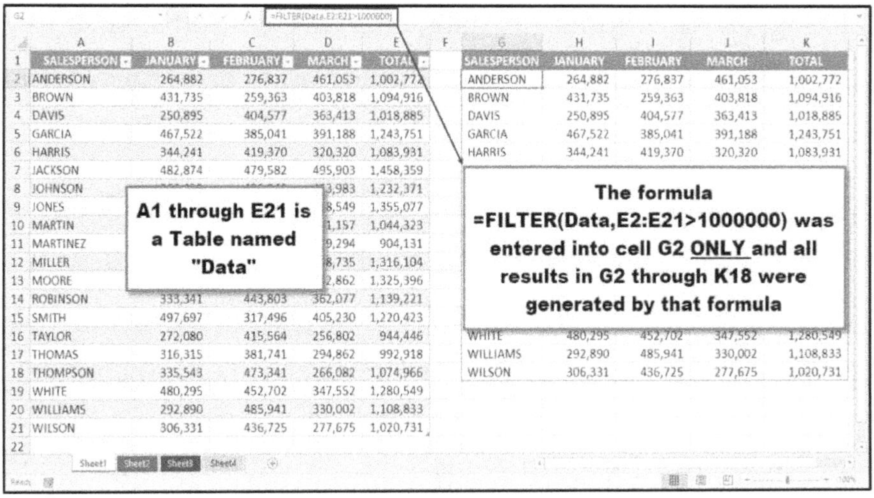

Figure - Using a FILTER to create Dynamic Array

The FILTER illustration presented can demonstrate the strength of the dynamic arrays as they allow you to evaluate data using formulas, with the calculation effects being related to but not affecting the original data set. As a result, you will run several forms of analyses on same underlying data set without copying it.

6.9 Utilize Power Query to evaluate the quality of the data

Power Query's evolution, which began with the 2010 introduction of Excel, is nothing short of amazing. This utility allows you to import data into Excel from various external data sources, such as the databases that serve the majority of large accounting applications. Even more specifically, Power Query enables you to convert the data to make it more usable. These transformations involve but are not limited to the following: deleting extra columns of data, merging columns, inserting user-defined calculations, and filtering and sorting as part of the query.

Power Query's capacity to automatically review the data for consistency problems such as completeness and precision is a recent enhancement. This functionality enables you to easily detect possible issues such as incorrect data, incomplete information or even duplicated records. To enable this function, check the Column quality, Column profile boxes and Column distribution on the View tab of the Power Query Editor, as shown in the Figure below. As you can see, Power Query creates a "quality snapshot" of each column of data in the query; however, clicking on any column in your query reveals a more comprehensive view of the data, providing details for that column and a graph of the column's value distribution. This function becomes accessible to Office 365 users in February 2020.

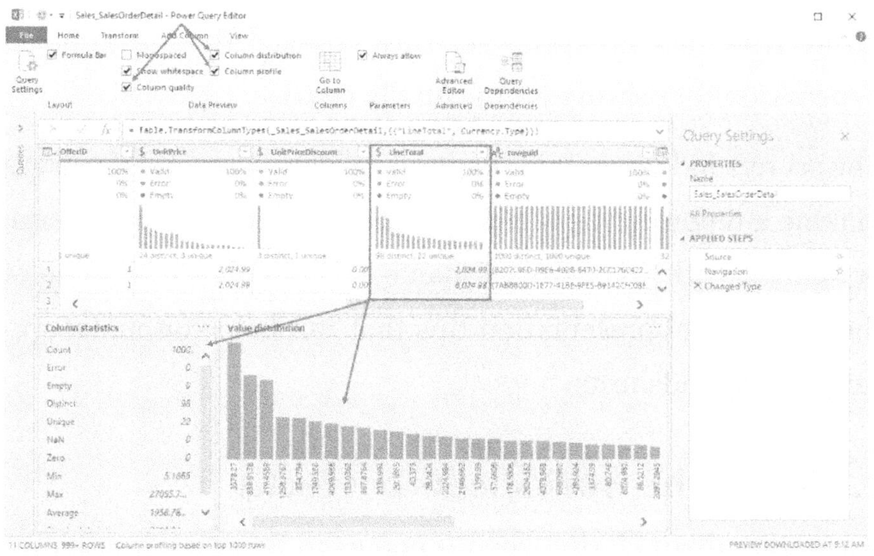

Figure - Using Power Query to analyze the data quality

Chapter No: 7 Some Excel Presentation tips and Professional skills

A spreadsheet is more than just a set of numbers on a page when it comes to Excel. Making the spreadsheets look professional, easy to understand, and visually pleasing to the viewers is equally important. Your Excel layout will not impress the users if it appears rough and boring, no matter how many hours of analysis went into it or how relevant the knowledge stored inside it is, much as a lawyer with crooked tie and jumbled papers would raise an eyebrow in the court.

The secrets hidden in this chapter can come in useful if you are making a report for personal use, passing details to your team, or sharing with your project manager. Let's glance at some of the latest Excel presentation tips that can help you make eye-catching spreadsheets.

7.1 Name the worksheets correctly

It is more about clarity when it comes to Excel presentations. The significance of a right and accurate worksheet or project name cannot be overstated for this single purpose. It may be an expression, a sentence, or just a single letter. Only make sure it is easy to grasp for you and everyone else with whom you'll be sharing the file.

You must also ensure that it is separate from the titles of the other worksheets on your device. After all, what use are all the lessons you are going to practice today if you cannot locate the worksheet you applied them on?

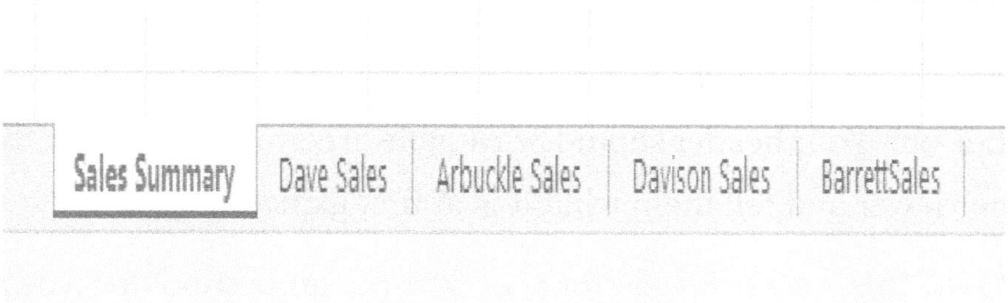

7.2 Get a template online

You may use a template to generate a new workbook in Excel. A template is a pre-designed spreadsheet. Excel comes preloaded with several templates, although some can be found on Office.com. Join the ex Machina: the pre-made Excel templates, whether you are a busy individual who cannot fit an Excel presentation concept into your timetable. You may choose from a variety of purpose-specific templates that include attractive styles, fonts, and colors. To customize it, simply enter your values, and you are ready to go. For instance, utilizing a template prevents you from improving your design skills. If having something accomplished is more important to you than getting great at making presentations, then go ahead and use a template.

7.3 Define your header or title

The information that exists at the top of each printed page is referred to as a header, and the information shown at the bottom of printed page is called to as a footer. New workbooks are usually produced without the footers and headers by default. Anything goes with your header and title, but it must stick out. Your header should be capable of communicating with the viewer and tell them what it is at first glance.

To do this, use a bigger font, underline, and embolden your header. It should be focused and in separate font color. It must stand out while still blending in with the template's color scheme and overall aesthetic appearance. For your header, you may also use a different readable. Only keep in mind that it to be distinct, not isolated.

	A	B	C	D	E	F	G	H	I
1	**Sales Report Summary**								
2	Widget Sales, Inc.								
3	For the period ending 1/1/2018 to 12/31/2018								
4									
5									

7.4 Font do's and do nots

Full transparency: The spreadsheet's fonts can make or break it. Often use a consistent font for your data; you may either use the same font for the header or modify it. You should only use three fonts in a single presentation, and that is the suggested maximum; otherwise, you would be overextending it. Less is still best in this situation.

These are the rules to follow when choosing the correct font format.

1. **Font size**

Since it necessarily depends on the presentation, font 12 with double spacing is often recommended to increase readability. As previously mentioned, the header font may be made larger. The headers should be bigger than the sub headers, which should be bigger than the data fonts.

DO:	Calibri
DON'T:	Calibri

2. **Font type**

If readability is a priority for you, sans-serif fonts are the right option for your Excel spreadsheet. Helvetica, Calibri, Arial, and Play fair are only a few examples of fonts. They will bring out the best of the Excel presentation if used with the correct orientation, spacing, and color.

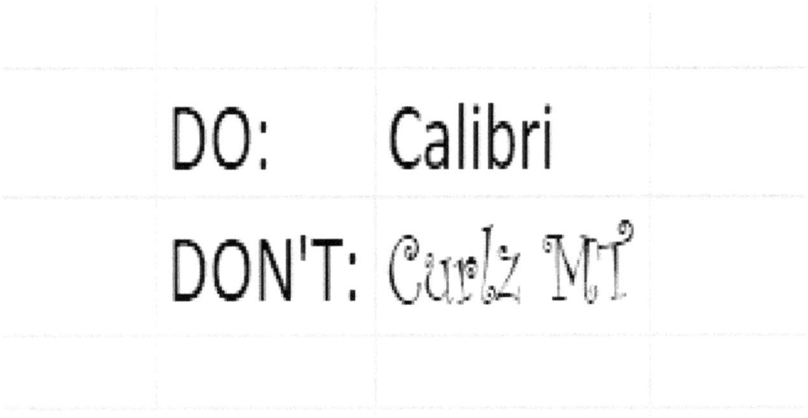

DO:	Calibri
DON'T:	Curlz MT

3. Font color

There is such a thing as seeing so much color in your life. Although this is often stressed in the fashion industry, it is often valid with Excel presentations. For the presentation, no more than two matching colors, colors in the same shade, or two contrasting colors can be included. You want to contrast the text color with the background color, for example, a bright color text on your dark background and vice versa. The "zebra stripes" are also included in this.

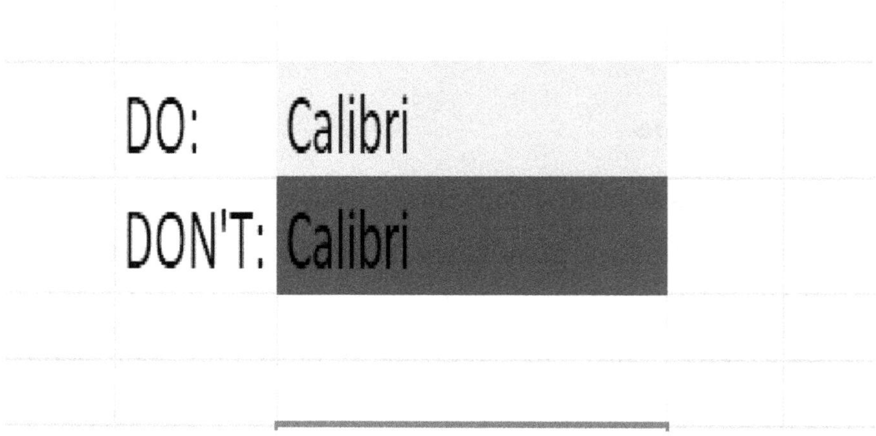

DO:	Calibri
DON'T:	Calibri

4. **Alignment**

People do not often use the alignment tool in Excel. If you want to make your presentation look beautiful and business-like, you will need to maximize the alignment feature. Headers should usually have center alignment unless it is better at the side. The data must have an extreme right alignment for the numerical or numbers data and a hard left alignment for the texts. The Center alignment is usually not advisable in your data input. To wrap the title or data around the cell, click on a cell and then go to Home toolbar, select crooked tie and disorganized papers Alignment and then select wrap text.

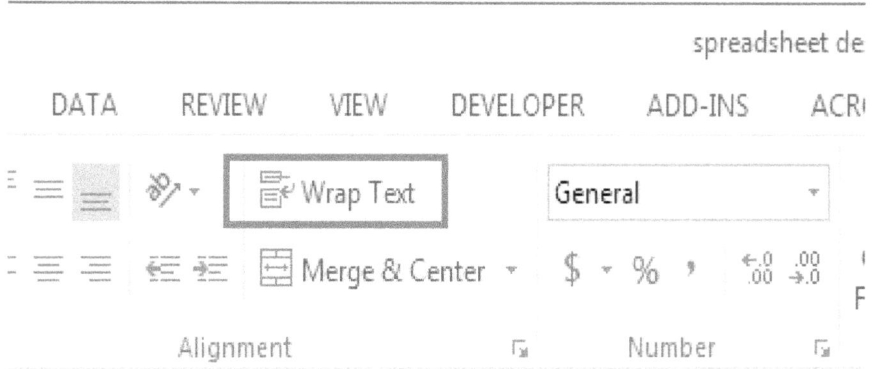

7.5 Create some space for breathing room

When you see a densely packed, wordy or sloppy text or spreadsheet, your brain gets bored reading it even before you begin. However, as there is breathing room and the spreadsheet is separated into categories, it becomes more appealing to the eyes and easier for the brain to interpret.

The B2 rule is the result of this. Begin the presentation on row 2 of column B. The A column and the first row are all blanks. It acts like magic. You can also double-check that the row and column dimensions are the same. Also, do not let the document's height and width auto fit. Your workspace must enable you to be flexible and innovative. Instead, manually change the height and width of the presentation so that it has only enough white space but not too much, giving it more breathing room and improving readability.

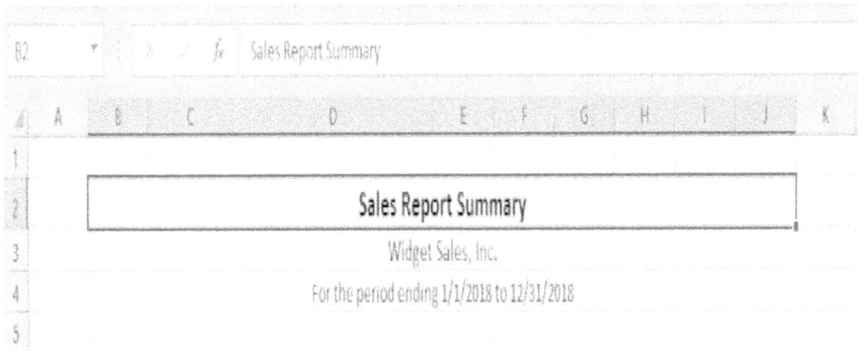

7.6 Add an image

If it is a photograph, an abstract drawing, or a slogan, pictures contribute significantly to the quality of your spreadsheet. Images give the presentation an official appearance and a professional feel, as shown in all of the amazing presentations you've seen. A picture is worth a thousand sentences. Although Excel is not intended to provide the same presentation as PowerPoint, using a picture can help you make your point and make the presentation iconic.

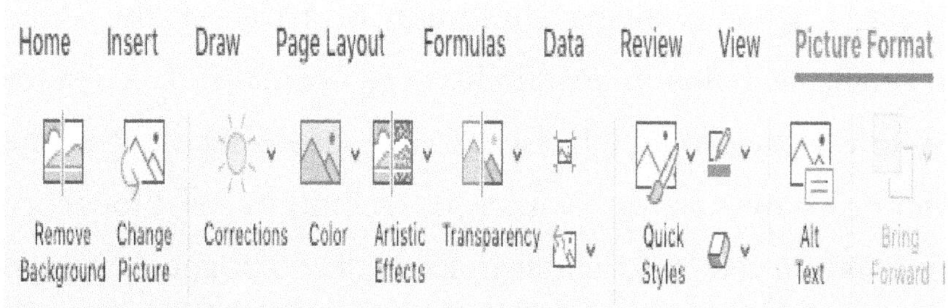

7.7 Go off the grid

Are you aware that by erasing all grid lines except those associated with your result, you can evoke questions about how you did it and whether you used the same Excel version as other people do? Consider that today.

1. Choose the View tab on the ribbon on the spreadsheet.
2. Uncheck the box next to the Gridlines in the Show section.

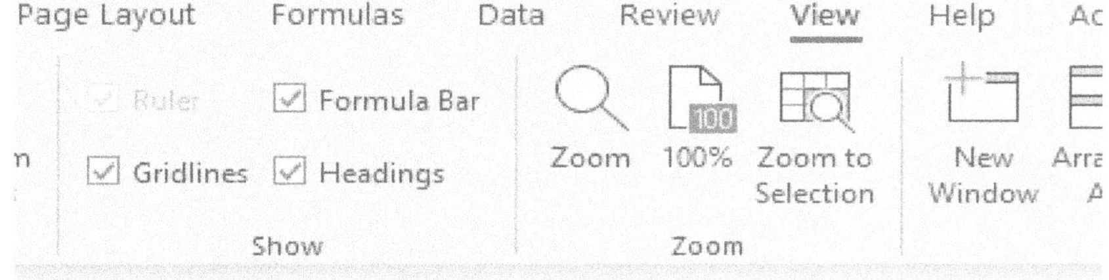

7.8 The Zebra stripes: Microsoft Excel jungle law

Zebra stripes are made up of bands of contrasting dark and light shades. This is beneficial in a variety of respects. First, it has an artistic quality that lends a sense of order to your work, particularly when viewing hundreds of rows of the data. Additionally, it helps in readability and correlation. A reader will follow a row from right to left without losing sight of the row her or his eyes are on. Numerous techniques exist for zebra striping. When you build a table in Excel, it will be zebra-striped by design (Tip- pick the data and easily create a table using the shortcut Ctrl + T on the PC or + T on a Mac). You can customize the color and pattern of the zebra stripes on the Design tab under the Table Styles.

If required, this may also be accomplished using a formula of conditional formatting. Conditional formatting is accomplished by emphasizing values that meet defined criteria (e.g., all the odd-numbered rows). The painter tool in the Home toolbar can be used to copy it from cell to cell.

Sales	2011	2012	2013	2014	2015	2016	2017
Arbuckle	$119,031	$128,673	$146,043	$161,816	$183,499	$182,398	$197,537
Barrett	$65,875	$69,498	$75,266	$83,019	$93,562	$101,609	$113,192
Davison	$120,201	$133,784	$151,978	$165,200	$169,165	$188,281	$196,189
Elliott	$75,440	$74,912	$74,163	$75,572	$77,008	$83,476	$82,558
Fletcher	$86,625	$95,894	$95,223	$108,744	$114,399	$131,559	$141,163
Herrman	$112,681	$110,991	$111,435	$125,141	$123,890	$125,872	$130,278
James	$100,065	$107,170	$116,279	$115,116	$120,642	$135,239	$139,567
Kent	$99,204	$110,116	$116,393	$124,308	$140,219	$148,492	$162,450
Kirov	$95,900	$107,504	$118,147	$124,290	$138,708	$137,182	$152,958
Myers	$88,765	$88,321	$100,774	$101,782	$102,495	$110,182	$123,183
Okonye	$121,027	$123,084	$122,346	$125,772	$144,637	$143,480	$158,833
Salgado	$60,149	$60,510	$66,379	$67,840	$71,164	$78,138	$85,483
Davison	**$120,201**	**$133,784**	**$151,978**	**$165,200**	**$169,165**	**$188,281**	**$196,189**

7.9 Use tables, charts, and graphs

Without any visual representation, the majority of presentations are incomplete. Whether in the form of a graph, table, or chart, you must visually reflect the raw data in formats easily interpreted at a glance. Charts, diagrams, and tables cannot be overlooked, even more so if the data is lengthy and spans several columns and rows. The graph, chart, and table features of Excel are like symbiotic siblings. You require them to carry out the elegance of your work's brevity.

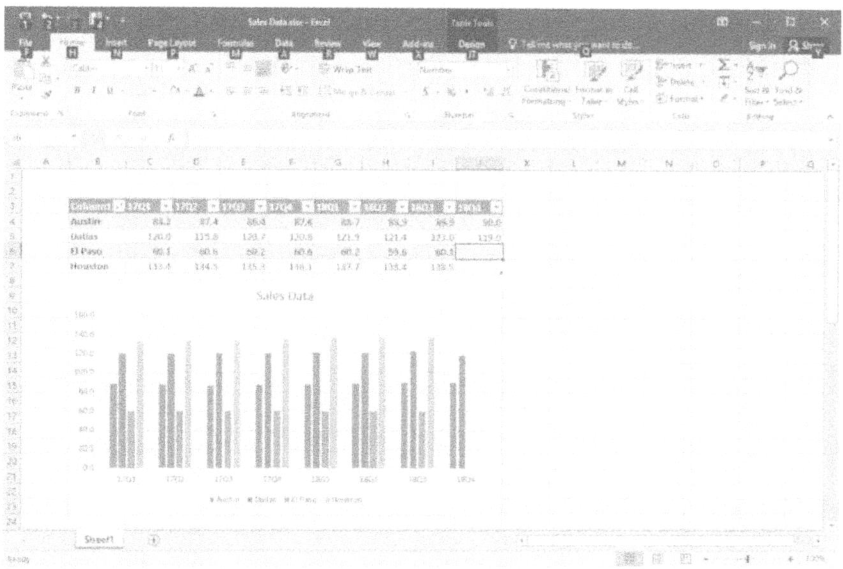

7.10 Create cell styles

Excel has various predefined cell types, but you may also build your custom styles that are more personalized and simpler to use and edit due to their creation. This is a viable alternative to purchasing a template if maintaining graphic accuracy is a priority. After making an elegant spreadsheet with the details above, you can save the theme for potential presentations.

Your presentation is now flawless, with the appropriate feel and style. Simply select the cells that contain your template for saving, then navigate to the Home toolbar, press "more" at the bottom of the style gallery, and then select "new cell style." A dialogue box for naming the style, editing its assets, and saving will open. If everything is not broken and functions well, why alter it? However, you can inject some variety into the design by adjusting the color palette from time to time.

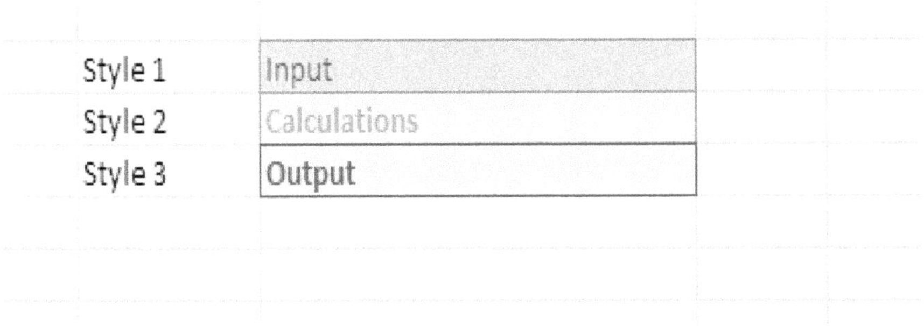

Style 1	Input
Style 2	Calculations
Style 3	Output

7.11 Show restraint

You've mastered every one of these tips and are ready to begin your presentation - but be cautious not to overdo it. Make sparing use of color and avoid combining so many suggestions at once. You must walk a delicate line between underwhelming and excessive to achieve the "just enough" balance. As in all, ensure that the presentation is perfectly balanced.

Finally, the quality of your Excel presentation is determined by how well you convey the data to the audience, while it is beneficial to understand the psychology of colors and decent fonts. Browse online for visually appealing spreadsheet displays to get a sense of what the "latest" feels like. However, the ball is ultimately in your hands, and the commitment to training, sharpening, and perfecting your Excel presenting abilities will be met with applause if you follow all these instructions.

Some Microsoft Excel skills:

You are still wondering which excel skills to learn first when there are so many. This chapter has compiled a list of the top 20 must-have MS excel skills for technical purposes to make it easier for you. For administrative programming jobs, you'll require the following simple Excel skills. These are the abilities that the majority of recruitment agencies look for in an applicant. Therefore, develop the following skills to please recruiters and reach the job sector with eloquence and courage and learn more about some of the Excel skills.

20 MUST-HAVE EXCEL SKILLS FOR PROFESSIONALS

1. DATA FILTERS
2. DATA SORTING
3. PIVOT TABLES
4. SUMIF/SUMIFS
5. COUNTIF/COUNTIFS
6. EXCEL SHORTCUTS
7. CHARTS
8. CELL FORMATTING
9. MANAGING PAGE LAYOUT
10. DATA VALIDATION
11. WORKBOOK
12. VLOOKUP
13. PIVOT CHARTS
14. FLASH FILL
15. QUICK ANALYSIS
16. POWER VIEW
17. CONDITIONAL FORMATTING
18. MOVING COLUMNS INTO ROWS
19. IF FORMULAS
20. AUDITING FORMULAS

7.12 Data Filters

This may have seemed to be quite a basic excel skill to some, but if you are unfamiliar with data filters, you'll need to develop this ability to differentiate yourself from the competition. When you understand how to use data filters, you can quickly sort, search and hide, critical pieces of information in a spreadsheet.

7.13 Data Sorting

Excel enables you to organize the data in the spreadsheets. For instance, you can sort in alphabetical or reverse alphabetical order. This is a difficult strategy to learn, and in some cases, you can unintentionally sort one column or row but not another, resulting in a spreadsheet being messed up. These tricky strategies are precisely why it is best to enroll in a well-structured Excel course.

7.14 Pivot Tables

These are summary tables that allow you to sum, count and average data in addition to performing other calculations. When you understand how to do this, crunching data and creating reports for your business becomes much simpler.

7.15 SUMIF or SUMIFS

This useful little Excel skin contains a feature that sums cells that satisfy various yet distinct parameters. It is often used when the adjacent cells follow text, date, or numeric requirements. In other words, it is a method for effectively organizing data on a spreadsheet.

7.16 COUNTIF or COUNTIFS

COUNTIFS is a feature in Microsoft Excel that counts the number of cells in a set that follows one or more requirements. COUNTIFS may be used as a worksheet function by using it in a formula in a worksheet cell. Though some of the Excel skills can seem unfamiliar at first, they become readily apparent as you pursue a formal learning path.

7.17 Excel shortcut keys

While dealing with Excel, learning how to use the shortcut keys will help you complete tasks even faster. Some of these keys can be accessed directly from the keyboard rather than from the mouse. While this shortcut just saves a few seconds every time you use it, if you deal with spreadsheets all day, you'll save a lot of time by mastering the shortcut keys.

7.18 Charts

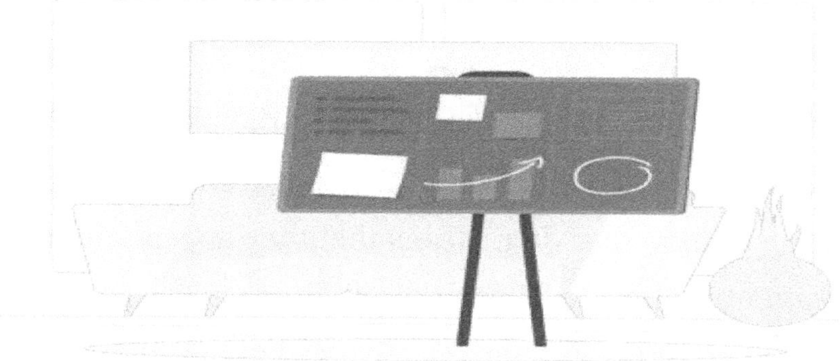

Another necessary skill is producing charts, which are excellent ways to display the data gathered by Vlookup. A chart is a graphical representation of data that you can use in Excel. Charts help the viewer's understand the context behind the figures and make it far simpler to demonstrate trends and comparisons. Considering the prevailing focus on data, it is unsurprising that the ability to produce charts is a widely sought-after Excel skill.

7.19 Cell Formatting

It is important to format data in a cell correctly so that it does not interfere with calculations or formatting. By formatting cells in Excel, we may alter the display of a number without altering the value of the number. It is a very fundamental Excel skill, but one that you'll need regularly. When a person learns how to format the cells in bulk, he will save a lot of time and seem more efficient and in command.

7.20 Managing Page Layout

Related to Cell Formatting, manipulating the page layout is a valuable Excel skill to possess. The page structure of the spreadsheets and other data means that they appear and print just as you want them to. Having the ability to navigate the web layout quickly helps you seem effective and organized.

7.21 Data validation

By using data validation, Excel will limit data entry to specific cells, ask users to enter the valid data when a cell is chosen and show an error text when a user enters the invalid data i-e Data Validation is used to limit the kinds of data that may be entered into cells. For instance, there might be a numerical limit to prevent you from entering incorrect figures, or the date may fall within a specified time frame. This is a useful tool for many applications that will assist you in maintaining leverage of any spreadsheets you build or work with.

7.22 Workbook

A workbook is a consolidated series of spreadsheets stored inside a single file. It is a really convenient way to have all of your linked spreadsheets in one place. It is another fundamental Excel skill that will be needed for every entry-level role in 2021.

7.23 Vlookup

VLOOKUP is an abbreviation for 'Vertical Lookup.' It is a feature that instructs Excel to look for a particular value in a column (the 'table array') to restore a value from another column in the same row. Vlookup is a necessary ability for someone who works with Excel. Vlookup enables you to consolidate data from various workbooks and sheets into a centralized location that is ideal for generating reports. This is a vital Excel ability for any analyst, that is a popular career right now, but it is also useful in general.

7.24 Pivot charts

Pivot Charts are a variation of the Excel chart. It enables you to digest complicated data by simplifying it easily. Pivot maps are very close to standard charts. They include data series and categories, but they vary in that they have interactive filters that enable you to access additional information easily.

7.25 Flash fill

The ability to fill out data efficiently rather than individually is referred to as flash fill. This conserves time and frustration. This might not sound like anything, but if you work full-time as an Analyst, for example, you will spend a significant amount of time dealing with numbers. This skill would save you a lot of time.

7.26 Quick analysis

This essential Excel ability is also a time-saver; it helps you create quick data sets, which decreases the time taken to create charts. You need to remember that it saves time!

7.27 Power view

Power View is a supercharged data analysis and visualization tool. It extracts and analyses massive amounts of data from external files and is an extremely useful and necessary Excel skill to have in 2021. Power view enables the creation of immersive, presentation-ready reports that can be exported to Microsoft PowerPoint.

7.28 Conditional formatting

Understanding how to use Excel's conditional formatting function enables you to classify data points of interest quickly. There are various rules and limitless implementations for this feature, making it essential to compete with the hottest 2021 excel skills.

7.29 Moving columns into rows

It is a normal phenomenon while dealing with data in Excel that you are working with columns but needs it in rows, or vice versa. Understanding how to do this simple trick is another time-saving Excel trick that you'll need in 2021.

7.30 IF formulas

IFERROR and IF are two very valuable Excel IF formulas. They allow you to use conditional formulas to compute in one way when a condition is valid and in another when it is false. For instance, you can classify 70 points or greater test scores by labeling the cell report "Pass" if the column C score is greater than 70 and "Fail" if the column C score is 69 or below.

7.31 Auditing formulas

Excel's formula auditing feature visualizes the interaction between cells and formulas graphically, enabling you to inspect the formulas and search for mistakes or edits. This functionality simplifies the process of auditing formula dependents and precedents, even object-based formulas. It has various applications in the workplace, and although it is a more technical skill, it is simple to pick up with the proper training.

Chapter No: 8 How to create Budget in Excel?

Let's face it; wealth is the driving force behind the culture in which we work. We attend school to obtain a respectable profession, and we invest in industry and other similar activities to earn income. Much of our efforts will be in vain if we are unable to handle our resources effectively. The rest of the people waste their earnings. To be financially successful, one must develop the habit of spending less than one receives and investing the remainder in profitable business ventures. So, this chapter will assist you in creating your budget strategy using Excel.

8.1 The fundamental components of a personal finance system

Since this is a fundamental personal finance system, so we'll examine the following components:

1. **Projected profits:** This is the sum of money you envision earning in the short and long term.

2. **Budget:** It contains a list of the items you want to buy, their numbers, and their associated costs.

3. **Actual salary:** which is the money you make consistently.

4. **Actual spending:** the amount of money expended on purchases.

The difference between estimated and actual profits is a success metric that indicates the accuracy of the forecasts or the efficiency with which a person works. The disparity between budgeted and actual expenditure acts as a success metric for our commitment to budgeting. Given that saving is intended to establish a personal financial plan relating real savings to actual spending monthly gives an estimate of how much a person will save over a year.

8.2 Excel spreadsheets for tracking income and expenses and creating personal budgets

You've studied the components of a personal finance system; now it is time to bring what you've learned into effect. You'll create two workbooks with this guide: one for revenue and one for budgets. Your workbooks should look like this at the end

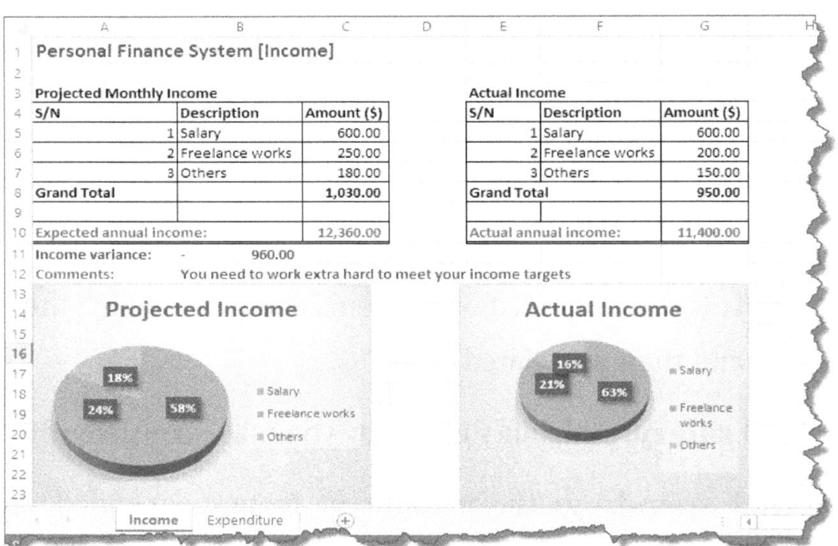

Personal Finance System [Income]

Projected Monthly Income

S/N	Description	Amount ($)
1	Salary	600.00
2	Freelance works	250.00
3	Others	180.00
Grand Total		1,030.00
Expected annual income:		12,360.00

Actual Income

S/N	Description	Amount ($)
1	Salary	600.00
2	Freelance works	200.00
3	Others	150.00
Grand Total		950.00
Actual annual income:		11,400.00

Income variance: - 960.00

Comments: You need to work extra hard to meet your income targets

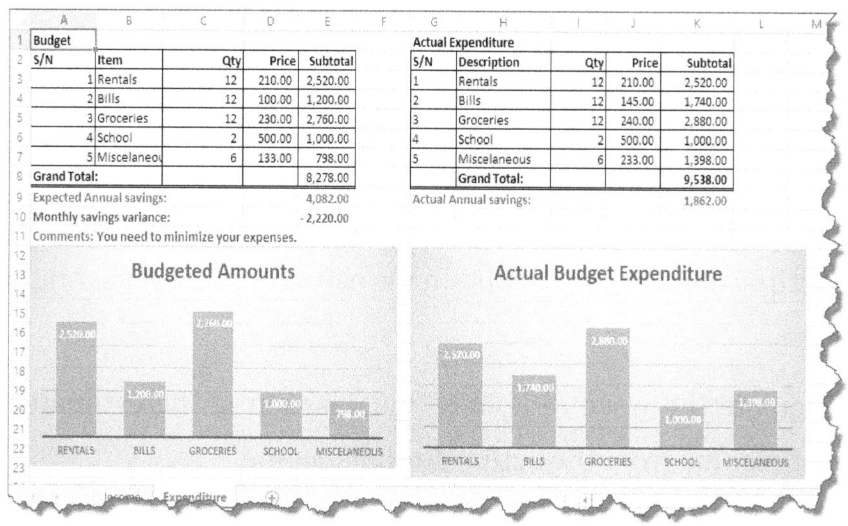

Budget

S/N	Item	Qty	Price	Subtotal
1	Rentals	12	210.00	2,520.00
2	Bills	12	100.00	1,200.00
3	Groceries	12	230.00	2,760.00
4	School	2	500.00	1,000.00
5	Miscelaneou	6	133.00	798.00
Grand Total:				8,278.00

Expected Annual savings: 4,082.00
Monthly savings variance: - 2,220.00
Comments: You need to minimize your expenses.

Actual Expenditure

S/N	Description	Qty	Price	Subtotal
1	Rentals	12	210.00	2,520.00
2	Bills	12	145.00	1,740.00
3	Groceries	12	240.00	2,880.00
4	School	2	500.00	1,000.00
5	Miscelaneous	6	233.00	1,398.00
	Grand Total:			9,538.00

Actual Annual savings: 1,862.00

8.3 Formulas for Income sheet

1. You must first determine your monthly revenue, both actual and projected.

2. Determine the average annual income and real annual income based on the monthly total.

3. Calculate the difference between the real and expected annual income.

4. Use analytical functions to help you understand what you are doing.

5. Make use of conditional formatting to draw attention to your financial reporting skill.

8.4 Exercise 1 of the Tutorial

1. Compile a list of all monthly income sources. Utilize the SUM command. This can be accomplished using both anticipated and actual monthly earnings.

2. Multiply the response to obtain the estimated and real annual revenue.

3. Calculate the wage difference by subtracting the estimated annual income from the actual annual income.

4. Insert a comments row directly under the columns. If the gap is greater than zero (0), use the IF functions to show "You must work especially hard to achieve your income targets," otherwise display "Excellent job working harder and smarter."

5. If the difference is smaller than zero, use conditional formatting to change the color of the text to red; otherwise, change the color of the text to grey.

8.5 Expenditure sheet

Create a new sheet and rename it Expenditure. In the sheet, enter the data in the format shown below.

Budget:

S/N	**Item**	**Quantit y**	**Cost**	**Subtota l**
1	Rents	16	540.00	

2	Bills	18	550.00
3	School	6	1000.00
4	Groceries	10	700.00
5	Miscellaneous	5	200.00
Grand Total:			

Actual income:

S/N	Description	Quantity	Cost	Subtotal
1	Rents	16	550.00	
2	Bills	18	590.00	
3	School	6	1000.00	
4	Groceries	10	650.00	
5	Miscellaneous	5	300.00	
	Grand Total:			

8.6 Formulas for Expenditure sheet

1. You must first determine the subtotal.
2. Using the subtotals, calculate the grand sum.
3. Determine the expected annual savings. The anticipated total costs were calculated as the difference between the estimated annual income and the total budget. Additionally, this will be done with real income and expenditure.
4. Evaluate the monthly savings difference.
5. Introduce a comments row directly under the columns. If the gap is less than zero (0), the message "You must limit your expenditures" is shown. Otherwise, view "You've done an admirable job of adhering to the budget."
6. If the difference is less than zero, use conditional formatting to change the color of the text to red; otherwise, change the color of the text to green.

8.7 Visualizing the data using charts

Charts are an outstanding method of visualizing data. Practice applying a pie chart to your salary sheet with the projected monthly income.

1. Your chart is limited to displaying details for the following columns:

- Column for the subtotal
- Column of items

2. Highlight items 1 to 5 in the Item section.
3. Select the subtotals from 1 to 5 in the subtotals list by pressing and holding the Ctrl key.
4. Select the INSERT button from the main menu.

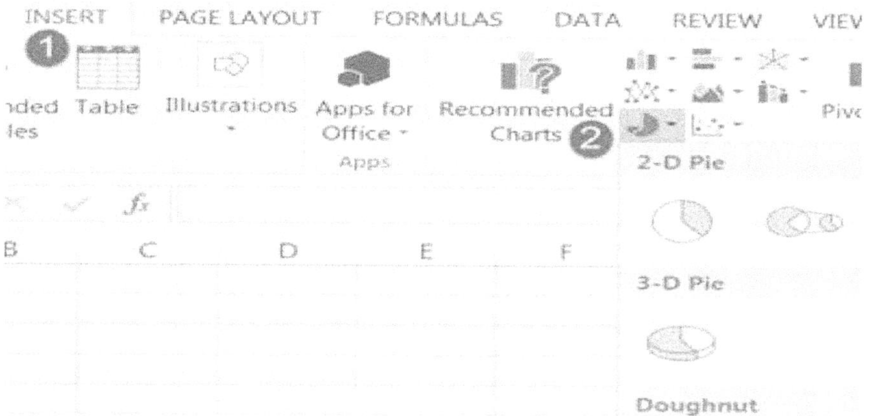

5. From the charts ribbon bar, choose a pie chart by tapping on the pie chart drop-down menu.
6. The pie chart can now look similar to the one below.

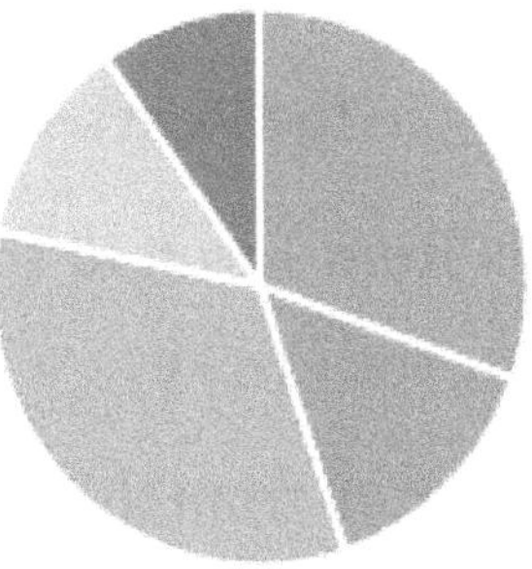
Chart Title
Rentals
Bills
Groceries
School
Miscelaneous

Chapter No: 9 Various Methods for Managing or Creating Workbooks/Worksheets

In Excel documents, the word "Worksheet" refers to a list of cells arranged in columns and rows. It is the work surface on which you interact to enter data. Each worksheet comprises 16384 columns and 1048576 rows and functions as a massive table for organizing data.

Inside Microsoft Excel, a workbook is a set of spreadsheets referred to as worksheets in a single file. Consider the picture below for an example of an excel spreadsheet. It is a component of a Microsoft Excel spreadsheet. This file is titled "Book1." This file contains two covers. The first sheet has been labeled "Sheet1," while the second has been called "Sheet2."

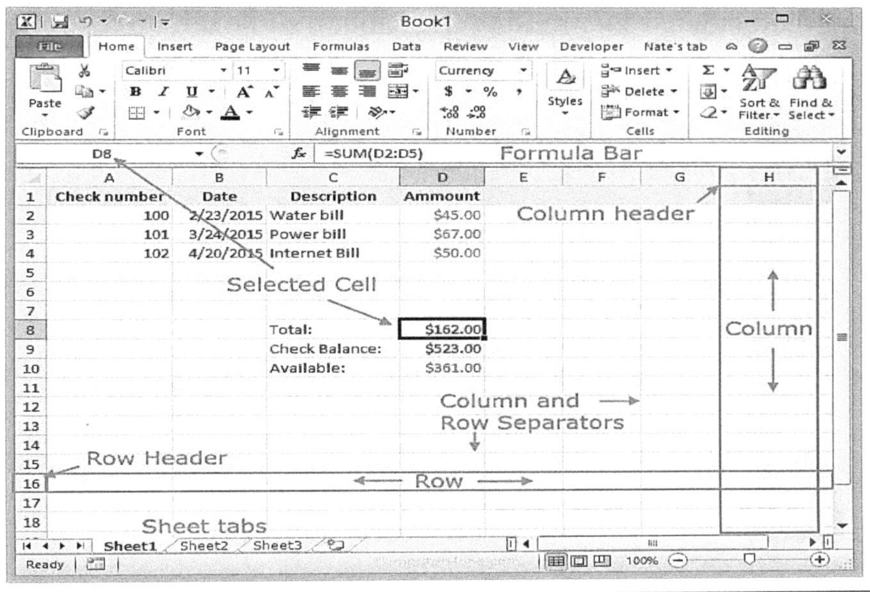

9.1 Design a blank workbook

This is almost certainly something you have done before. This is an easy task. When you open Microsoft Excel for the first time, a window will appear asking what a person wants to do.

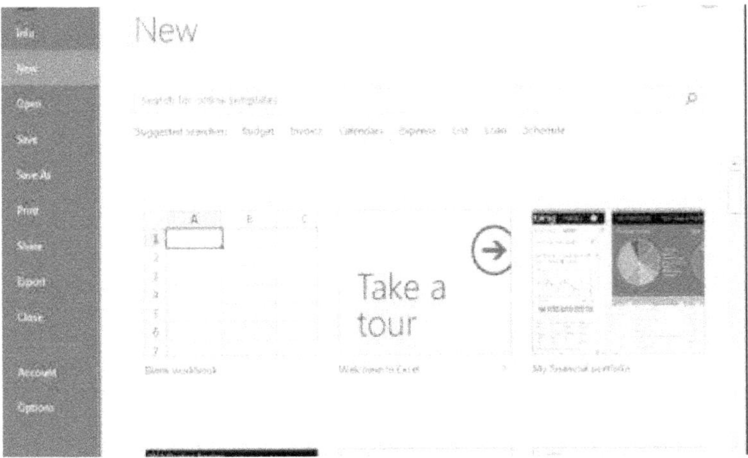

To open a new workbook, double-click on a blank workbook. If a workbook is accessible, choose the "new" option from its upper left sidebar by pressing the Ribbon's File tab's left spot.

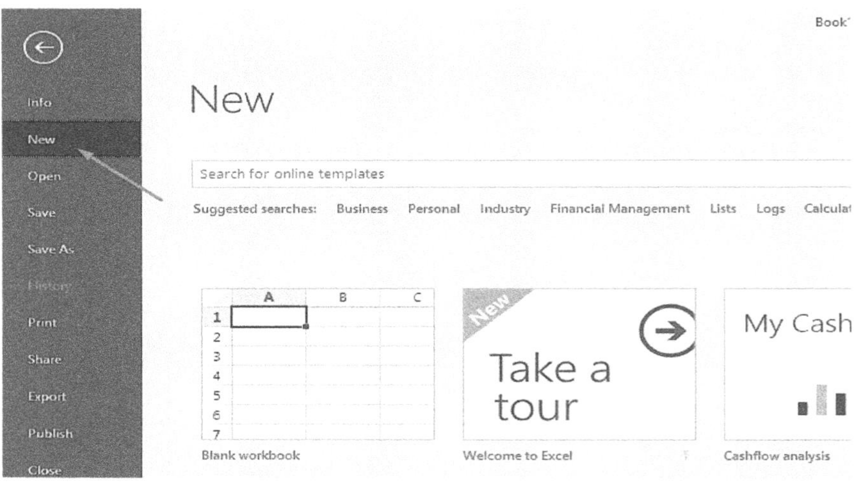

Simply double-click the new Workbook once more to complete the operation.

9.2 Using the excel system to design a template for opening a workbook

If this is your first time saving a workbook to a personal template, begin by creating a default location for personal templates:

1. From the File menu, select File > Options.
2. Select Save, followed by Save a Workbook.
3. The most common method is to go to C: Users, then to [UserName] Documents, and finally to Custom Office Templates.
4. Press the OK button.

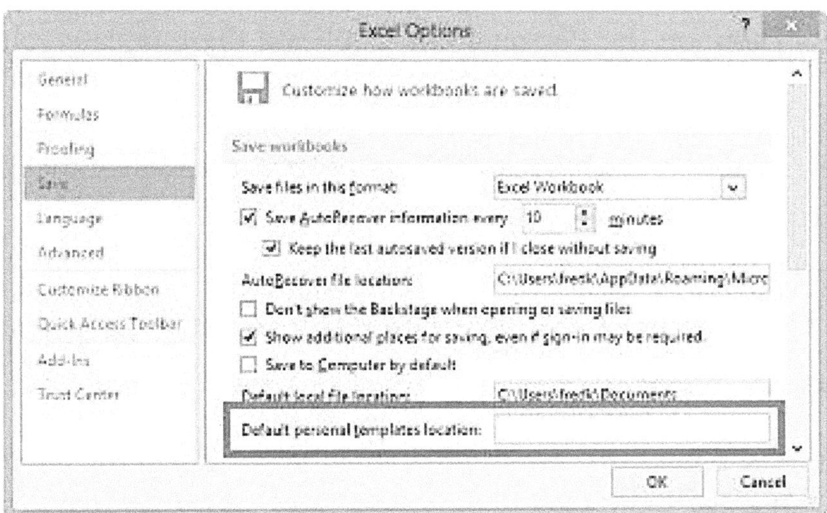

5. If this option is enabled, all custom templates stored in the My Templates file will appear in the Personal File section of the New page (File > New).

The following procedure describes how to navigate a workbook from a template:

1. From the File menu, choose File > New.

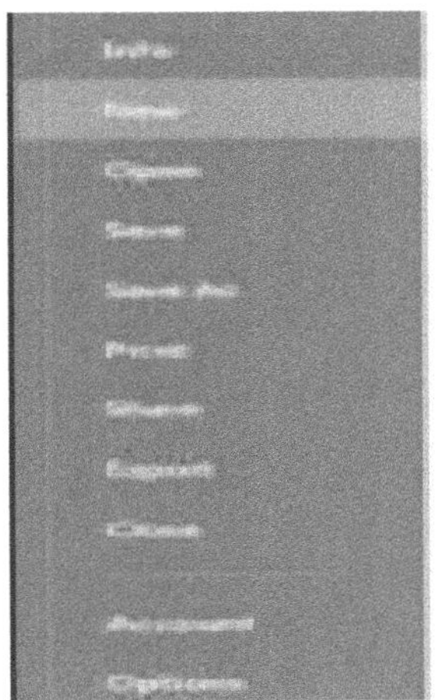

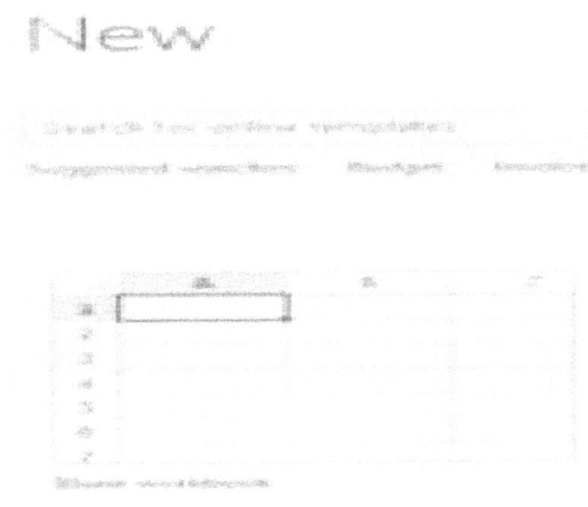

1. Then, on the Personal tab, duplicate the template you just created.
2. In this case, Microsoft Excel creates a new workbook based on the template.

9.3 Open a previously created workbook

To access a previously saved workbook in Microsoft Excel 2k21:

1. Press the FILE tab on the toolbar at the top of the main display.

2. Now, press the left-hand "open" tab. And, on the right, a chart of "Recent Workbooks" would appear. To quickly access these workbooks, tap on the Workbook's title under "Recent Workbooks."

3. Rather than using the "Recent Workbooks" tab, navigate the Computer tab and click the Browse button.

4. This will open a search window for the Workbook's Microsoft Excel file. After selecting the File to access, press the Open button.

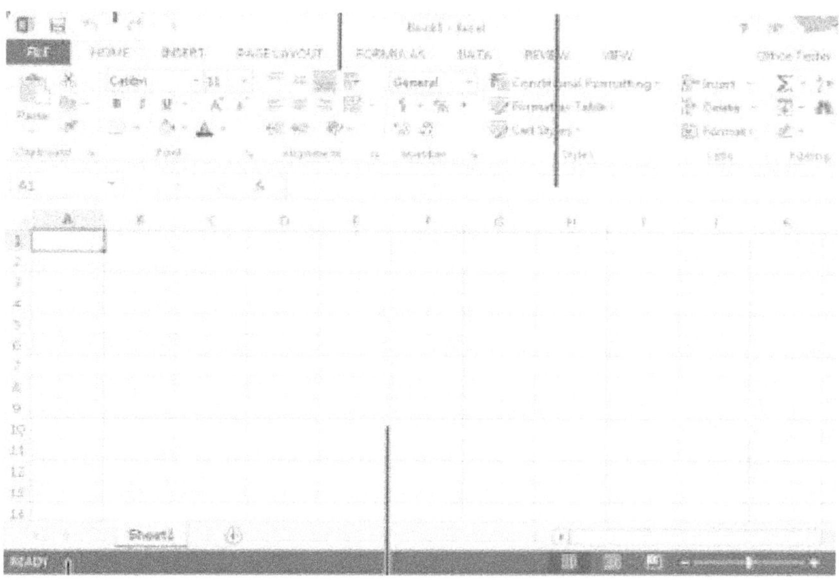

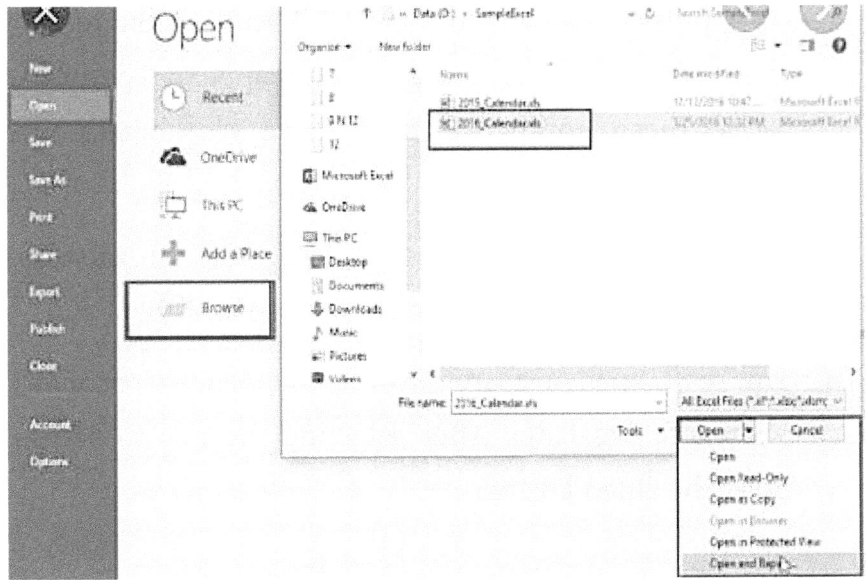

5. If you want to open a file instantly in Microsoft Excel, double-click on it; else, right-click on it to open the file.

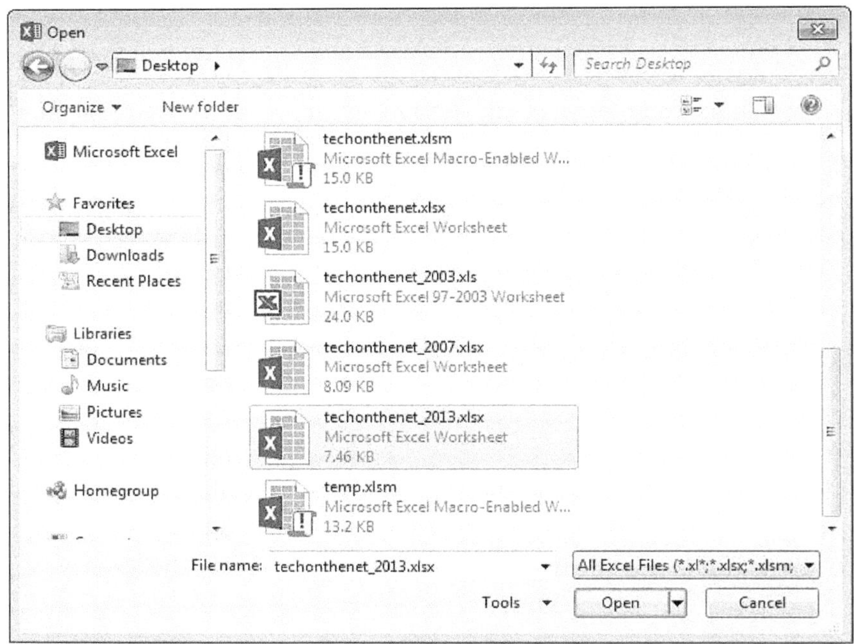

9.4 An illustration from the business world

There seems to be a sizable selection of free Microsoft Excel templates ready for download. Follow the measures below to create a new workbook file based on an Excel template. Inside Microsoft Excel 2k21,

1. Select the File tab, then New.
2. A list of MS Excel templates will appear.

The following formulas will be included in the Excel 2k10 version:

1. Choose from several example templates, which are pre-installed templates on your screen.
2. Browse the com Templates category, click on any category to see thumbnails, and then import the template of your choice.

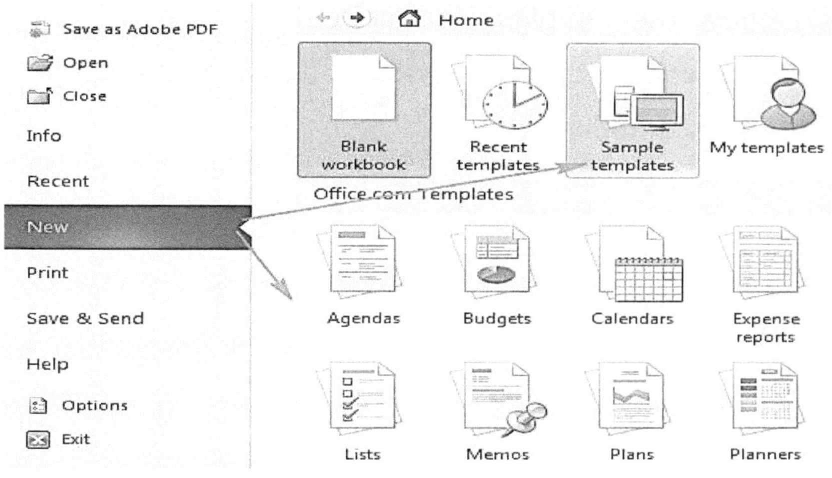

Simply click on any template to see a sample. A sample of the desired template will appear, along with the publisher's complete title, additional details, and a how-to guide.

Many that want to display the template before installation will do so by clicking the Create icon. As an example, suppose you've selected an appropriate calendar template through MS Excel:

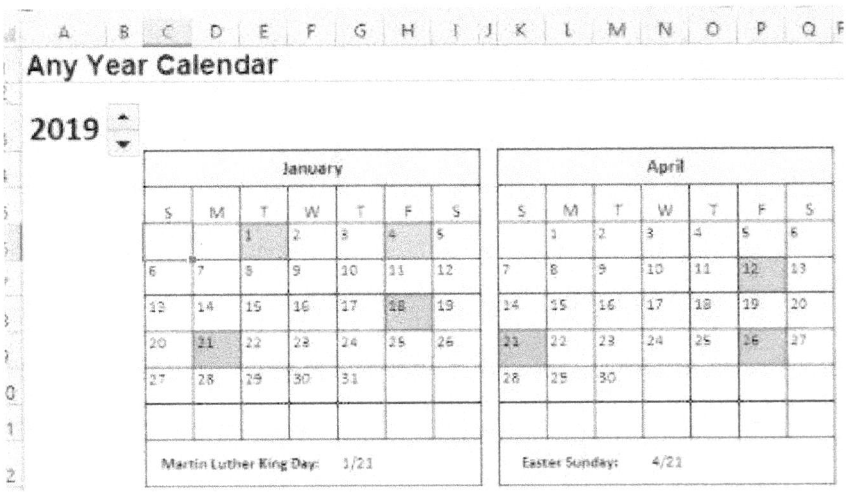

This is now the template you selected; it will be easily downloaded, and an updated workbook based on this template will be created automatically.

9.5 Native or non-native Excel files

A native file appears to be the directory structure of the electronic file, as specified by the excel program. For example, if an Excel spreadsheet is created, the native layout of the document will be the Microsoft Excel (.xls) format. Additionally, non-native files may be enabled in Microsoft Excel. To open a non-native file in Micro Soft Word 2k21's File Explorer, right-click the file, select Open With, and select Word (on the desktop).

To enable the software window in Word 2k13 to open some non-native file, follow these steps:

1. Navigate to the required file location in the backstage display using the already-opened tab.
2. Inside the dialogue box, choose all files to view all files included inside a folder, or select the file form you are looking for on the right side of the File screen.
3. Select the relevant File to open from the Open dialogue window, then tap the open button.

9.6 Linking Excel to the External Sources

Microsoft Excel enables users to import data from a range of sources.

- Access to databases
- The internet pages

- Documents in Text Format
- Workbooks from the other sources

Now, you will get to know how to import data quickly and easily from an external file. You may easily import data from other folders that classify them as text files if the data is included in a CSV file, text file, or text. To do this, take the following steps: -

1. To begin, create a customized Microsoft Excel Worksheet.
2. Following that, navigate to the Ribbon bar and choose the Data tab.
3. Select from the text to access the Outer Data Group. The Import File dialogue box then appears.

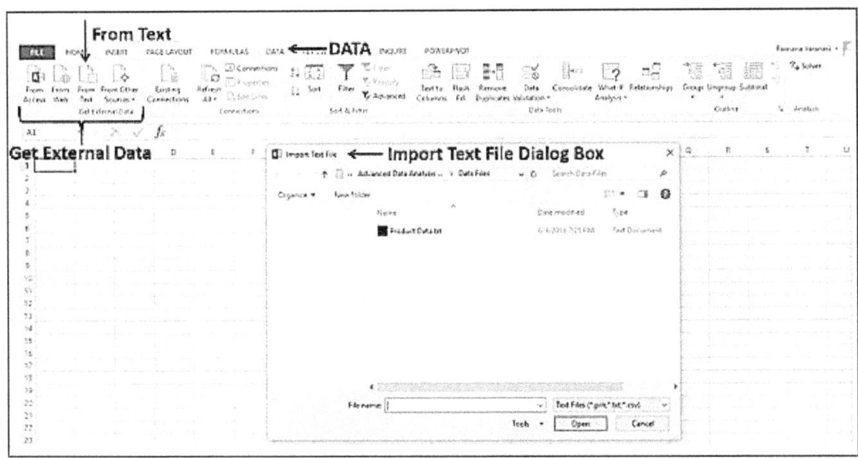

You'll now see a list of the text, print, and CSV file extensions that have been approved.

- Select a specific file.

- This File Name box includes the file's name.
- The open key has been repurposed as the Import button.

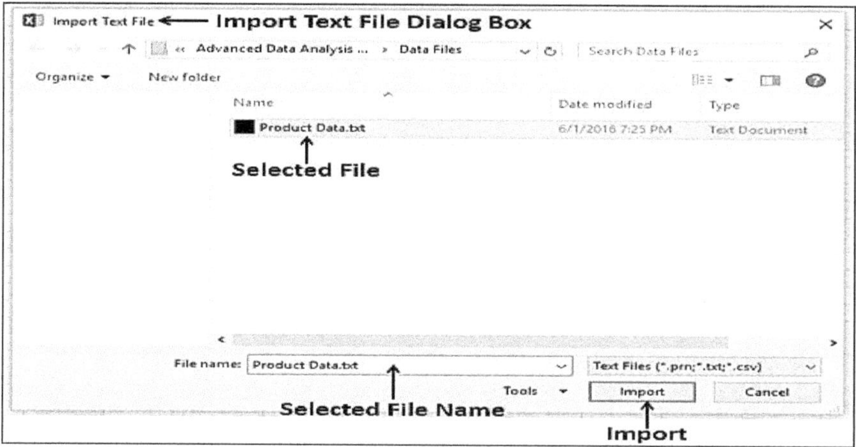

- From the drop-down menu, choose Import.
- Following that, the Text Import Wizard tab would appear.
- Select the File alternative from the Delimited list, and then tap next.
- Select other from the drop-down menu for Delimiters.

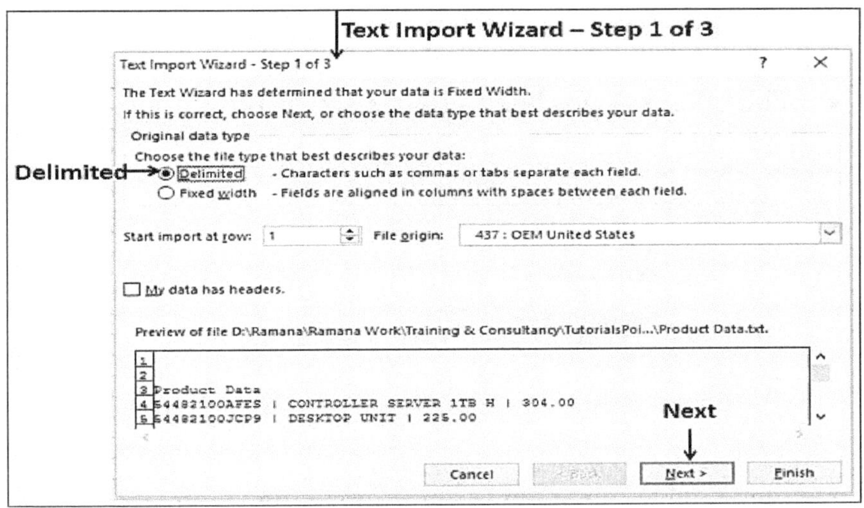

- Simply type in the adjacent box, followed by the next key.

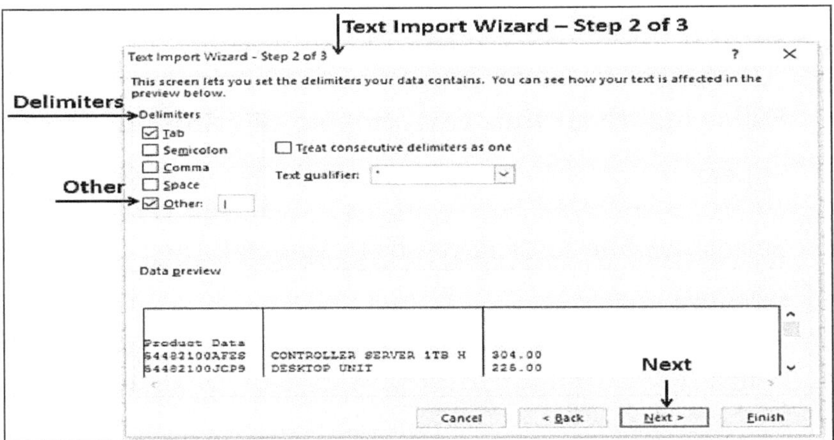

- You will customize the specifics of each column's format in this dialogue view.

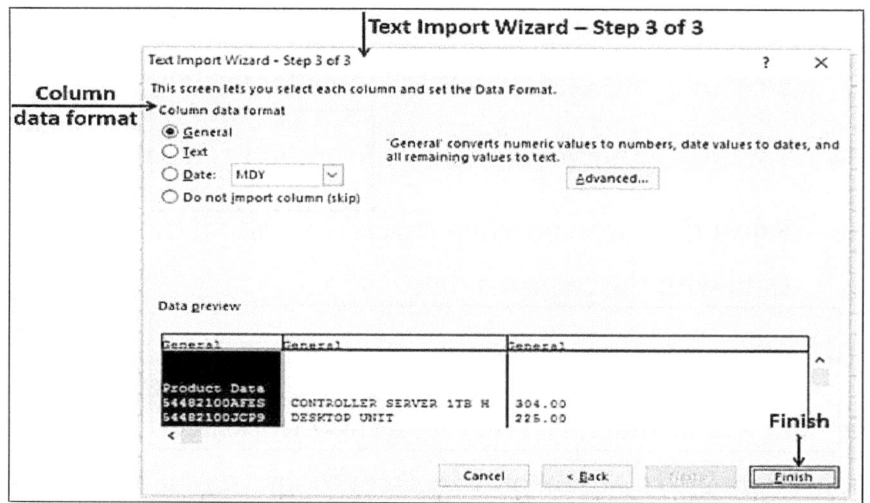

- If you've completed editing the data columns, press finish to open the Import Data dialogue panel.

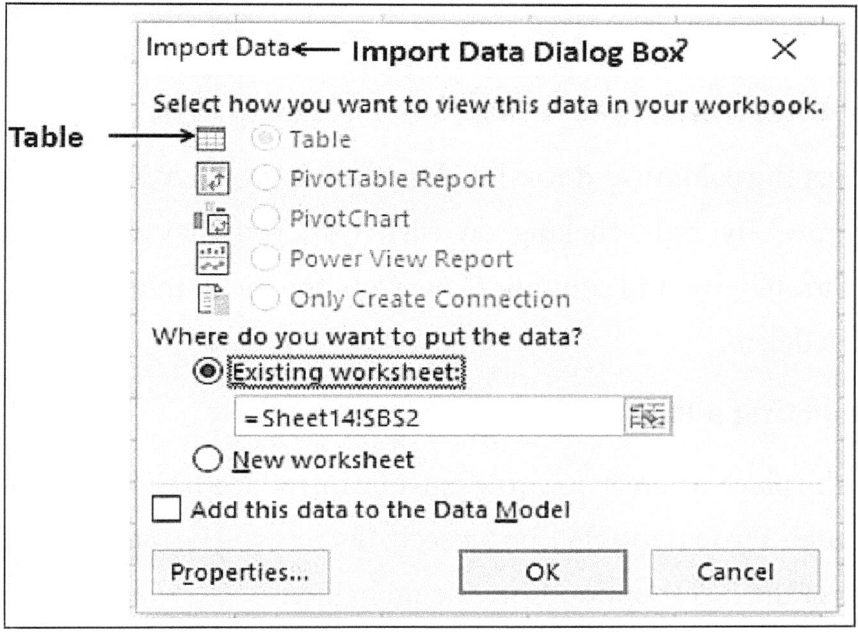

Consider the following example:

- For selected, a greyed-out table appears. In this case, you have only one option: use the table for viewing.
- You may either create a new file or modify a current one.
- Select the corresponding check box and fill the Model of Data with the required data.
- When done, press OK.

The data will be displayed on the desired worksheet. You must import data into a Microsoft Excel workbook from a Text file.

9.7 Worksheet operations

There is a list of many common Excel worksheet operations that beginners and small business owners can consider.

Eliminating Columns and Rows from a Table

Select the column and row headers to remove an entire column or row. By right-clicking on each cell, you may create an illustrated row and column. Then, from the menu that appears, click delete.

Inserting a Row

At the place where a new row must be introduced, click the row header. If you right-click on any cell, the row will be highlighted. Select the insert option from the main menu.

Inserting a Column

At the place where a new column must be added, click the column header. If you right-click on any cell, the column will be highlighted. Select the insert option from the main menu.

Sorting

To choose the whole worksheet, press the grey rectangle in the middle, which includes the "A" column header and the "1" row header on the Excel worksheet's upper left hand. Collect the results using the menu bar, then sort... Select the column to filter by and specify whether to sort descending or ascending.

Show Formulas in a Worksheet

Long-press the CTRL button, then tap the left quotation key (which is usually denoted by a "tilde" ()). Rep the procedure to revert to its numeric View. Note: You can print the data from this spreadsheet in both the Formula and Numeric displays.

Formulas

Select the cell where you want to insert the Formula and then type it in. Take note that all formulas must start with an equal sign (=), by tapping on the equal sign (=) in the formula bar, you may obtain additional details about the functions.

To resize a column

Simply tap a column header to select all items. Using the right-click menu, select any cell inside the highlighted column. Column Width can be selected from the drop-down menu to... Again, from the main menu, enter the appropriate column width value. Take note that by choosing several columns at once, you can modify the widths of several columns simultaneously.

Height adjustment for the rows

To highlight the whole line, click the row header. Right-click on every cell inside the highlighted row. Then, from the drop-down menu, choose the Row Height icon... From the main menu, enter the value required for row height adjustment. Note that

various row heights can be changed concurrently when selecting several rows.

Altering the numerical data format

Select and copy the cells containing numeric details. Simply right-click on any cell to access the highlighted section. Select the way to format cells. To specify the form of data contained inside cells, navigate to the Number section and choose Category. To adjust the decimal places in numerical data, choose a number and enter the decimal places to use.

Justification of cell contents

Just choose the correct cell. So, choose the Justification indicator from the Formatting Toolbar (right middle or left justify). Keep in mind that the bold, italicized, and outlined buttons on the right side of the window represent all three keys.

Rows and columns justification

To highlight an entire column or row, simply click on the column or row headings. Select the Justification option from the Formatting Toolbar (either center, left, or right justify). Keep in mind that these three keys are located in the right corner of the bold, italics, and underline controls.

All cells, columns, or rows can be copied

You will prefer the column(s), cell(s), and row(s) by highlighting them. Simply right-click on every highlighted cell to bring up the menu bar. To use the Copy tab, click it. The cells that have been chosen would be copied to the Windows clipboard (invisibly). Keep in mind that Windows only retains the most recent copy folders.

Pasting cells, columns, and sections are all possible:

Choose the column(s), cell(s), and row(s) onto which the freshly copied information is to be pasted (i.e., of Windows clipboard). Keep in mind that the region selected would be the same size as the data you paste. Right-click on any colored cell. Following that, from the drop-down display, choose the Paste button.

The laws of arithmetic precedence

As Microsoft Excel does formula checking, it often refers to the arithmetic precedence rules.

() Procedures enclosed in parentheses are evaluated first, accompanied by nested parentheses, which are analyzed from the inside out.

- ^ Exponentiation
- and / A method of multiplying and dividing that is evaluated from left to right
- + and - are left-to-right addition and subtraction operators.
- and / A process of multiplying and dividing that is evaluated from left to right.

As an example

A formula =10*(6-8) 4 is evaluated like 5.

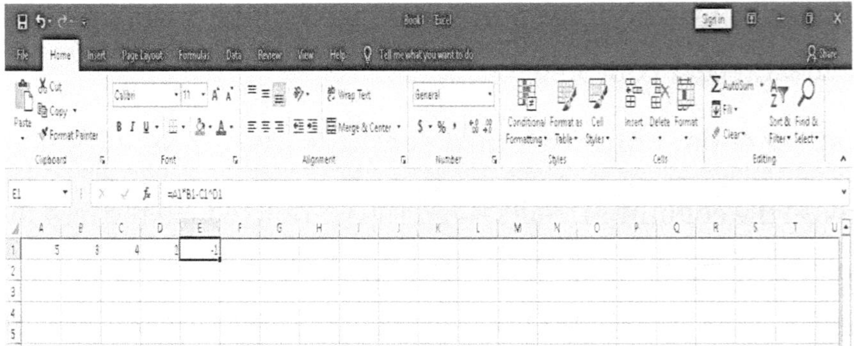

9.8 Excel worksheet operations

To adjust the color of a Worksheet section in Microsoft Excel, press the Worksheet section you want to edit.

1. In the Ribbon, click the "Home" icon.
2. Now, press the "Format" key on the "Cells" segment key.
3. Browse down to the "tab color" control.
4. From the main menu bar, choose the color you want to add to the worksheet page.
5. Other than that, press the "More Colors..." list icon to bring up the "Colors" dialogue panel, where you may choose a color.
6. Alternatively, click "Nil Color" from the menu to clear the chosen worksheet page's color.
7. Deselect the selected worksheet segment to see a more detailed view of the variety of colors you've created.

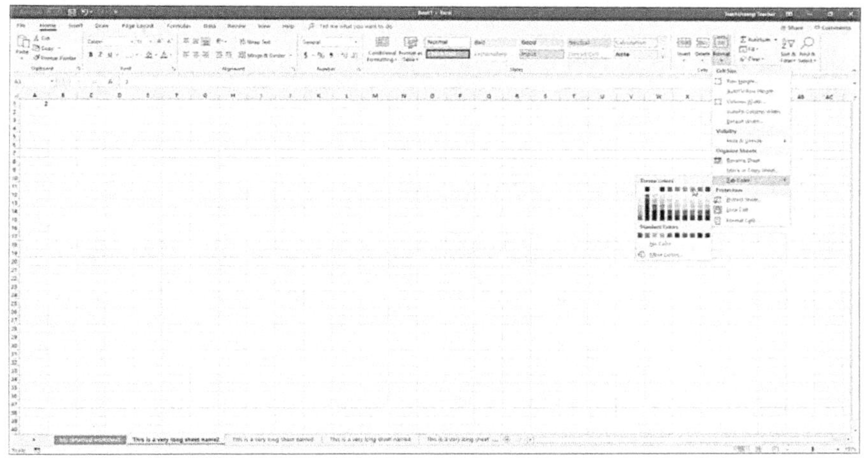

Conclusion

Microsoft Excel is the most well-known and commonly used spreadsheet program in the world. Created by Microsoft in 1987, this software has seen several enhancements through the years, making it the go-to source for the spreadsheet editing, graphing applications, pivot tables, and macro programming, among other things. This software is also available on all digital browsers, including Windows, Android, macOS, and iOS, and is used daily by hundreds of millions of users. Since 1990, Microsoft Excel has been included in the Microsoft Office suite, a set of document, presentation, and email editing tools that cover many of the use scenarios needed in a modern collaborative work setting. The new Excel versions have everything necessary to get started and then develop into a specialist and other excellent features. Microsoft 2021 introduces a slew of innovative updates and improvements, the most notable of which is the fact that Excel will be open to you as part of their subscription package for the first time. Thus, while changes occurred every few years in the past, you can now anticipate them daily.

MS Excel understands and organizes trends and data, thus saving you time. Create spreadsheets quickly and easily using templates or by scratch, and then conduct calculations using modern features. Via Microsoft 365 on Excel, you can share workbooks with others while continuing to work on the fresh versions, with real-time synchronization allowing you to complete tasks rapidly. It includes both fundamental and specialized technologies that can be found in almost every market environment. The Excel spreadsheet enables you to create, display, edit, and exchange data with others rapidly and conveniently. You will build spreadsheets, data reports, data tables, and budgets when reading and updating excel files attached to emails. If you gain more familiarity with various definitions, you can know the latest tools and features that Excel offers its users. The fact is that you can fulfill almost every personal or company's wish by using Excel's functionality. What you have to do is devote your time and skills. Although the learning curve for certain talents would be lengthy, you will see that certain abilities become second nature through experience and time. After all, it is by practice that a man becomes better. At the end of the day, no other tech rival has such a consistent experience and many optional features. So Microsoft excel is best suited for you to do your work efficiently.

www.ingramcontent.com/pod-product-compliance
Ingram Content Group UK Ltd.
Pitfield, Milton Keynes, MK11 3LW, UK
UKHW021911190726
13853UKWH00002B/615